S0-AWV-417

Developments in Marine Corrosion

Developments in Marine Corrosion

Edited by

S.A. Campbell, N. Campbell and F.C. Walsh

Applied Electrochemistry Group, University of Portsmouth, Portsmouth, UK

The Proceedings of the 9th International Congress on Marine Corrosion and Fouling, held at the University of Portsmouth on the 17–21 July 1995.

Special Publication No. 177

ISBN 0-85404-763-8

A catalogue record for this book is available from the British Library

Published by The Royal Society of Chemistry,
Thomas Graham House, Science Park, Milton Road,
Cambridge CB4 0WF, UK

For further information see our web site at www.rsc.org

Printed by Bookcraft (Bath) Ltd.

Preface

The 13 papers collected in this book are taken from the 9th International Congress on Marine Fouling and Corrosion held at the University of Portsmouth, 17-21 July, 1995.

The diversity of subjects involved in marine corrosion and fouling processes are evident in the approaches and techniques described by the international authors. An improved understanding of biologically influenced corrosion and fouling processes is essential to the development of better corrosion protection systems and the papers here show new perspectives in our appreciation of these subjects. Equally, powerful microscopic imaging, surface spectroscopy and electrochemical techniques have become available and are providing new information on metal surfaces experiencing fouling or microbial corrosion. Examples include confocal laser microscopy, scanning probe microscopy, Fourier Transform infrared spectroscopy, scanning electrochemical microscopy and scanning (or vibrating) reference electrode techniques.

Advances in our knowledge of biocorrosion processes have enabled improvements in corrosion monitoring and protection strategies. Examples include metal or polymer coatings and the use of corrosion inhibitors and biocides.

The international authors are to congratulated for their contributions and it is hoped this book will provide a valuable contribution to the literature on biocorrosion in the marine environment.

Sheelagh Campbell and Frank Walsh

Preface

The [illegible] papers [illegible] International Congress on [illegible] Metal Forming and [illegible] at the University of [illegible] 17–21 July [illegible]

[illegible] the diversity of [illegible] advances in [illegible] of metal [illegible] processes [illegible] and approaches [illegible] constitutive [illegible] and [illegible] processes [illegible] surfaces [illegible]

[illegible] of [illegible] processes [illegible] and [illegible] Examples include [illegible]

The [illegible] for their contribution [illegible] of [illegible]

[illegible]

Contents

List of Contributors

H.A. Videla
Bioelectrochemistry Section, INIFTA University of La Plata, CC 16. Suc. 4.1900, La Plata, Argentina.

B.J. Little & P.A. Wagner
Naval Research Laboratory, Stennis Space Center, MS 39529-5004, USA.

G.W.J. Radford, F.C. Walsh J.R. Smith, C.D.S. Tuck & S.A. Campbell
Applied Electrochemistry Group, School of Pharmacy and Biomedical Sciences, University of Portsmouth, St Michael's Building, White Swan Road, Portsmouth PO1 2DT, United Kingdom.

J.F. Halsall, M. Kalaji & L.K. Warden-Owen
Department of Chemistry, University of Wales at Bangor LL57 2UW, Wales, United Kingdom.

G. Denuault, L. Andrews, S. Maguire & S. Nugues
Department of Chemistry, University of Southampton, Highfield, Southampton SO17 4BJ, United Kingdom.

D. Féron & I. Dupont
CEA-CEREM, Service de la Corrosion, d'Electrochimie et de Chimie des Fluides, B.P. 6, 92 260 Fontenay-aux-Roses, France.

D. Wagner, H. Peinemann & H. Siedlarek
Institut für Instandhaltung GmbH, Kalkofen 4, 58638 Iserlohn, Germany.
Märkische Fachhochschule, Laboratory of Corrosion Protection, Frauenstuhlweg 31, 58644 Iserlohn, Germany,
Prymetall GmbH & Co. KG Zweifaller Strasse 130, 52224 Stohlberg, Germany

J.P. Busalmen, M.A. Frontini & S.R. de Sánchez
Corrosion Division, Instituto de Investigaciones y Tecnología de Materiales (INTEMA), Fac. De Ingeniería, University de Mar del Plata, Juan B. Justo 4302, 7600 Mar del Plata, Argentina.

Th.N. Skoulikidis, P. Vassilou & S. Vlachos
Department of Materials Science and Engineering, Faculty of Chemical Engineering, National Technical University of Athens, 9 Iroon Polytechniou Street, Athens 157 80, Greece.

G.S. Beloglasov & S.M. Beloglasov
Department of Chemistry, Perm State University, 15 Bukrievea ul., Perm 614600 and Department of Chemistry, Kaliningrad State University, 14 Alexander Nevski ul., Kaliningrad 236041, Russia.

P.L. Bonora, F. DeFlorian, L. Fedrizzi & S..Rossi
Laboratory of Electrochemistry, Materials Engineering Department, University of Trento, Messiano (TN), 38050, Italy.
AGIP Offshore, S. Donato (MI), 20097, Italy.

P.L..Bonora, S. Rossi, L. Benedetti & M. Draghetti
Materials Engineering Department, University of Trento, Messiano 77,38100 Trento, Italy.

N. Koulombi, G. Tsangaris & S. Kyvelidis
National Technical University, Chemical Engineering Department, Materials Science & Engineering Section, 9 Iroon Polytechniou Street, Athens 157 80, Greece.

1. BIOCORROSION PROBLEMS IN THE MARINE ENVIRONMENT: NEW PERSPECTIVES

Hector A. Videla

Bioelectrochemistry Section.
INIFTA University of La Plata.
CC 16. Suc. 4.1900. La Plata. ARGENTINA

1. INTRODUCTION

Microbial biofilms develop within a few hours on all metal surfaces exposed to natural marine environments. At long exposure times, marine fouling occurs in the form of a complex community of plants, animals and microorganisms that considerably alter and influence the immediate surroundings of the metal surface.

Biofouling and biofilm formation are the result of an accumulation process which is not necessarily uniform in time or space[1]. Thus, a new "biologically conditioned" metal/solution interface is formed and there is a reciprocal conditioning between the passive layers and the biofilms. Electrochemical concepts and measurements used for assessing inorganic corrosion in seawater in the absence of biofilms need to be revised and adapted to correspond to the characteristics of the biofouled metal surface[2]. Consequently, a bioelectrochemical approach is required to understand the complex passive layers/biofilm. relationships developed at the metal/solution interface.

In an aggressive medium like seawater, metal dissolution takes place simultaneously with biofilm formation. Thus, a very active interaction between the corrosion process and biofouling settlement can be expected[3]. The consequent corrosion behaviour of the metal will vary according to the intensity and nature of this reciprocal interaction. Biofilms are influenced by both the substratum and the bulk phase. On an active metal like carbon steel, the gelatinous structure of the biofilm, essentially formed by a matrix of extracellular polymeric substances (EPS), bacterial cells and water, is mixed with corrosion products that were formed within the same time scale. The observation of bacterial cells using electron microscopy is difficult, and complex corrosion products/biofilm interactions develop. Conversely, on passive metals like stainless steel or titanium, the lack of corrosion products on the metal surfaces allows biofouling settlement over an almost uniform passive layer of oxide. In a few hours, copious microfouling deposits of bacteria, EPS, and particulate material are generally formed, leading to a patchy distribution of the biofilm.

The two main influences of biofilms on corrosion are contrasting, either retarding or accelerating metal dissolution[4]. Retardation or "polarisation" of the metal surface means a reduction in the metal reactivity and may be due to a "barrier effect" of the biofilm, as reported in the literatures. As biofilms are rarely uniform, the opposite effect of enhanced metal dissolution is prevalent as a consequence of the permanent separation of anodic and cathodic sites at the metal surface, the breakdown of inorganic passive films, or the stimulation of either the anodic or cathodic reactions.

It has been reported[5] that the interactions between biofilms and inorganic passive layers determine the passive or active behaviour of metal surfaces in biologically conditioned media. Metal surfaces of different reactivity in the marine environment will be chosen to illustrate their dissimilar biocorrosion/biofilm interactions: high corrosion resistant alloys like stainless steel and titanium, active metal surfaces such as carbon steel and intermediate behavior alloys like copper-nickel and brass.

Taking into account new perspectives in the study of biofilms and biocorrosion, a brief overview on cathodic protection and biofouling interactions, and updated strategies for monitoring biocorrosion in offshore structures and coastal power plants will be provided. Finally new insights into biofilm systems, derived from recent improvements in analytical, microbiological, electrochemical and microscopical instrumentation will be considered.

2. DIFFERENT TYPES OF BIOCORROSION AND BIOFOULING INTERACTIONS IN THE MARINE ENVIRONMENT

2.1 Corrosion-Resistant Metal Surfaces

Titanium offers an outstanding resistance to a wide variety of environments. Its corrosion resistance is due to the presence of a stable, protective, strongly adherent oxide film. This film forms instantly when a fresh surface is exposed to air or moisture. Several special characteristics that account for the use of titanium in seawater and other aggressive environments are: i) good resistance to chloride containing solutions, ii) good performance in the presence of hypochlorites and chlorine compounds, iii) good resistance to nitric acid solutions and iv) when alloyed with palladium shows passivity in hydrochloric acid[6]. Exposure of titanium for many years to depths of over a mile below the ocean surface has not produced any measurable corrosion[7]. Pitting and crevice corrosion are totally absent, even if marine deposits form. The presence of sulphides in seawater does not affect the resistance of titanium to corrosion. These characteristics can be ascribed to the very high anodic pitting and repassivation potentials that make the metal less susceptible to pitting corrosion, the most frequent type of attack in biocorrosion[8]. It has also been reported[9] that titanium is fully resistant to reduced chemical species generally related to anaerobic microbial activity, such as ammonia, sulphides, hydrogen sulphide, nitrites, ferrous ions, and organosulphur compounds, as well as to biogenic organic acids over a wide concentration range. Titanium does not display any toxicity towards marine organisms. Thus, biofouling can rapidly occur on surfaces immersed in seawater. It has been reported[10] that bacterial adhesion on titanium surfaces exposed to cultures of

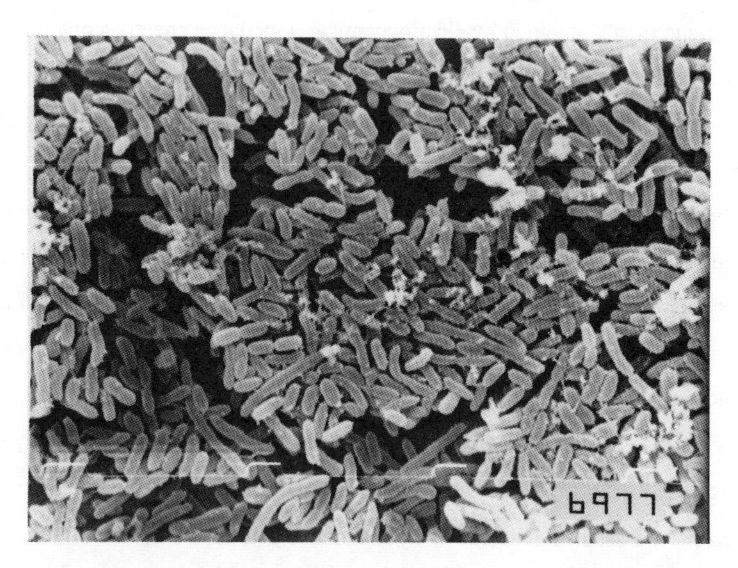

Figure 1 *SEM photograph of titanium surface, after immersion for 5 h in a* Vibrio aleinolyticus *culture (bar indicates 10 mm).*

Vibrio alginolyticus in artificial seawater for only 5 h, where a patchy biofilm has been reported (Figure 1). Although a decrease of the redox and corrosion potentials with time was found in these experiments, no evidence of corrosion attack was detected after 5 days exposure to bacterial cultures. Extensive biofouling on titanium after 800 hours immersion in shallow seawater has been reported[11]. The integrity of the corrosion resistant oxide film, however, was fully maintained under marine deposits and no pitting or crevice corrosion was observed.

Marine fouling on titanium heat exchanger surfaces can be minimised by maintaining water velocities in excess of 2 m s^{-1} [12]. Chlorination is recommended for protection of such surfaces from biofouling where seawater velocities less than 2 m s^{-1} are used.

An evaluation of titanium corrosion behaviour exposed to thermophilic and marine biofilms has recently been carried out[13]. After a one-year exposure period to Pacific Ocean seawater, grade 2 titanium was found to be colonised by diatoms and other algal species. No ennoblement of corrosion potential with time was found for titanium samples probably due to the poor catalytic activity of the oxide layer for oxygen reduction. Capacitance measurements obtained by electrochemical impedance spectroscopy remained constant showing values close to the theoretical ones observed for inert materials. After long-exposures of 1 year to seawater, titanium surfaces remained macroscopically clean, but covered by a thin biofilm of diatoms and microalgae.

A study of the marine biofouling characteristics for titanium tuber heat exchanger has been recently carried out simulating the expected service conditional. The fouling behaviour was seen to be dependent upon seawater temperature, tube wall heat transfer conditions, seawater flow velocity and time of the year as well as the site location. Continuous addition of chlorine or batch dosing of a quaternary phosphonium compound at 100 ppm v/v for 2 hour periods, every 168 hours were able to reduce biofouling by up to 80%. In this respect, titanium passive behaviour remains unaltered when strong oxidising biocides like ozone are used[15]. It seems likely that ozone acts as a stabiliser of the passive oxide layer of titanium due to its high oxidising power.

In summary, several and varied examples of titanium use in natural seawater have demonstrated its immunity to biocorrosion. However, its susceptibility to biofouling has been also pointed out. Consequently, its use must be controlled when the biological content of the intake water is high. Here, it has been shown[16] that, during the early stages of biofouling, the nature of the metal surface plays a relevant role in biofilm development, facilitating or hindering microbial adhesion. Corrosion-resistant materials such as titanium, present ideal substrata for microbial colonisation as was demonstrated after exposing several metal surfaces to flowing seawater containing high levels of pollution.

The amount of biomass deposited decreased in the following order: titanium> stainless steel>aluminium>brass>copper/nickel>copper. In spite of this affinity to biofouling, specific examples of titanium uses in natural seawater have demonstrated its immunity to biocorrosion.

As in the case of titanium, stainless steels present ideal substrata for microbial colonisation, due to the lack of corrosion products on the metal surface. Thus, biofilm formation and structure are more easily observed by electron microscopy. After several hours of exposure to natural seawater, a uniform layer of oxide covers the metal surface and pioneer bacteria begin to adhere by means of EPS production (Figure 2).

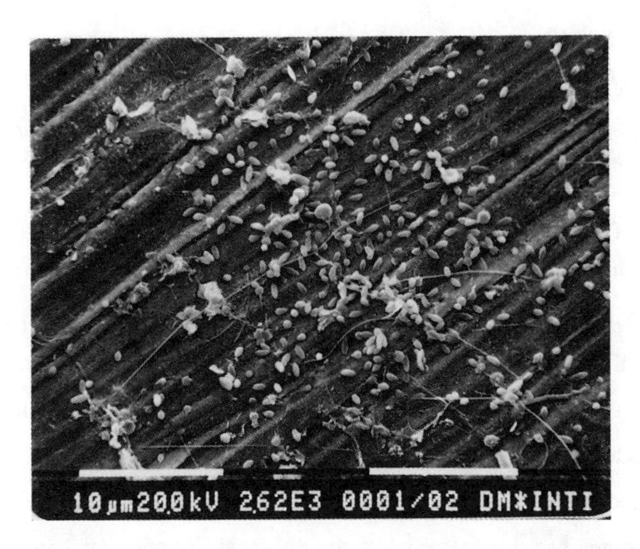

Figure 2 *SEM photograph of a stainless steel surface after several hours of immersion in flowing natural seawater (flow velocity of approximately 0.3* $m\ s^{-1}$*) showing the attachment of pioneer bacteria (the bar indicates 10 mm).*

After one week or more, complex microfouling deposits of bacteria, their EPS, particulate material and other large organisms (diatoms, protozoans, etc) can be found as a patchy distribution on the steel surface (Figure 3).

Figure 3 *SEM photograph of a stainless steel surface after three weeks of immersion in flowing seawater (flow velocity of approximately 0.3 m s^{-1}). A diverse microfouling can be seen on the metal surface (bar indicates 0. 1 mm).*

Biological deposits, and the metabolic activity of the microorganisms within the biofilms, can have an impact on the electrochemical reactions of the corrosion process. Microbial fouling favours the formation of localised anodes and cathodes, and hence corrosion, through a differential aeration cell mechanism. As the corrosion reaction progresses and the corrosion products accumulate on the metal surface, the biofilm/metal interface becomes gradually more heterogeneous. Biofilm accumulates forming a barrier for certain chemical species. For instance, microelectrode measurements in biofilms showed that dissolved oxygen had disappeared within a thickness of approximately 180 mm from the metal surface[17]. Previous conceptions of mass transport to biofilms were mainly based on diffusional processes, driven by concentration gradients across the system. Thus, if consumption of soluble substrates would be faster than their transport from the bulk solution, the system would be diffusion rate-limited[18]. Use of the Confocal Scanning Laser Microscope (CSLM) has provided high-resolution images of hydrated-living biofilms and has revealed a complex structure of microbial cell clusters and interstitial voids[19].

Within water channels convective flow will be predominant, and based on this finding, an entirely new scheme for oxygen distribution within the biofilm has been conceived: oxygen may be more concentrated in the void areas whereas bacterial cells and by products will be more concentrated under cell clusters. Thus, the local increase in oxygen concentration in the void areas will favour the formation of differential aeration cells and the initiation of corrosion. By measuring the current distribution with a Scanning Vibrating Microprobe (SVE) and imaging the biofilm with the CSLM, a relationship between biofilm heterogeneity and corrosion rate has been established[19].

The mechanisms of biocorrosion on stainless steels are mainly related to i) creation of differential aeration areas due to a patchy distribution of biofilms[4], ii) effect of biofilms on the onset of crevice corrosion through the depletion of oxygen in restricted areas[20] and iii) corrosion potential ennoblement with time[21,22]. Mechanism i) has been already mentioned in this section and mechanisms ii) and iii) will be discussed in more detail below.

2.2 Intermediate Corrosion Behaviour Metal Surfaces

Copper-based alloys such as (90:10 and 70:30) copper-nickels, admiralty brass and aluminium bronze are usually employed as heat exchanger materials due to their widely reported good anti-fouling properties. Notwithstanding this, it has been reported that after several months of exposure to natural seawater, a multi-layer structure of microorganisms and their EPS were found entrapped between layers of different inorganic corrosion products on copper-based surfaces[23].

It is well known that one of the mechanisms, by which biofilms participate in the corrosion process, is by chelation of certain metal ions by the EPS in the biofilm matrix[24]. Thus, the toxic effect that would be developed by cupric ions leached from the copper surface is counterbalanced by metal ions binding by the EPS. This layered biofilm/corrosion products structure is easily disturbed by shear stress effects arising from flowing seawater, which induces partial detachment of the biomass and corrosion products resulting in an increase in biofilm heterogeneity and hence differential aeration effects[25], (Figure 4).

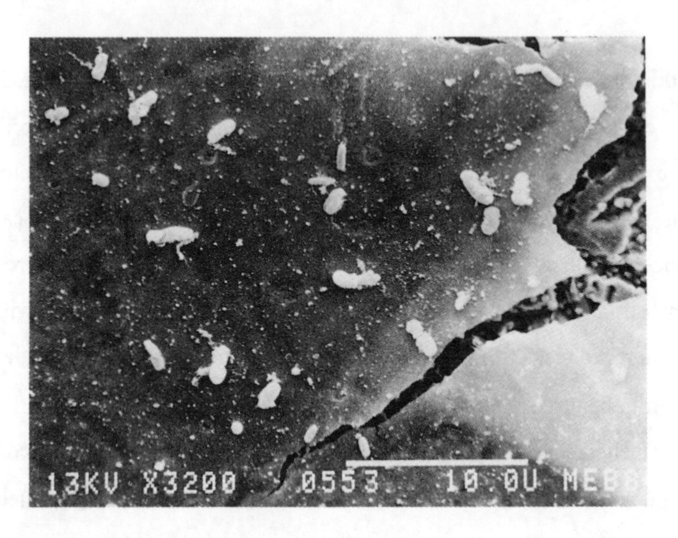

Figure 4 *SEM photograph of a 70:30 copper nickel surface after three weeks of exposure to flowing natural seawater (flow velocity approximately 0.3 m s^{-1}) showing the layered distribution of corrosion products and bacteria (bar indicates 10 mm).*

In laboratory experiments using sterile seawater, chloride ions can lead to an imperfect passive behaviour of copper alloys covered by certain types of corrosion products such as copper hydroxychloride. This compound usually replaces the uniform passive film of cuprous oxide on the metal surface by another porous and poorly passivating layer on copper-nickels surfaces exposed to artificial or natural seawater[16]. When significant levels of pollution in the seawater are present, several other factors may influence the corrosion resistance of copper-based alloys, including the attachment of microorganisms to the metal surface, the presence of a diverse range of sulphur anions (e.g., sulphides, bisulphides and hydrogen sulphide) related to the degree of pollution of the water or to the presence of sulphate reducing bacteria (SRB) [26]. The complex electrochemical behaviour for copper-based alloys in polluted seawater was found to be due to an alteration in the distribution and the structure of biofilms and passive layers on the metal surface. Generally, corrosion product layers formed in artificial seawater are more uniform and compact than those formed in natural seawater due to the adsorption of organics and biological material as well as to the precipitation of copper salts and sulphur compounds. Adhesion effects at the fixation points of organisms, e.g., *Zoothamnium sp.* in natural seawater, facilitate the detachment of the outer layers of the biofilm/corrosion products structure. Sulphur compounds, including sulphides, also play a role in corrosion by leading to the formation of an imperfect oxide layer having poor protective characteristics[27]. In these circumstances, copper nickel dissolution is higher than that observed in plain chloride solutions. Similar results were observed for aluminium-brass samples exposed to sulphide containing media[10]. Energy Dispersive X-ray analysis (EDX) of corrosion products formed on copper nickel alloys give further support to the role of sulphur compounds and SRB metabolises in the corrosion of copper based alloys in seawater. These effects are combined with differential aeration mainly in restricted areas beneath biological deposits. Differences in substratum composition can also influence colonisation by marine organisms. For instance, the effect of the iron content on the colonisation of 90:10 copper nickel alloy in seawater has been reported[28]. It was found that whereas iron-containing alloys were easily colonised, alloys without iron were rapidly corroded and scarcely colonised.

2.3. Active Corrosion Behaviour Metal Surface

Due to its wide industrial applications in the marine environment, carbon steel will be chosen to illustrate biocorrosion/biofouling interactions in seawater. The corrosion behaviour of mild steel in saline media is markedly conditioned by abundant deposits of corrosion products of varied chemical composition. For carbon steel samples exposed to natural seawater, external layers of corrosion products containing lepidocrite (γ-Fe_2O_3.H_2O) and wurtzite (FeO) has been reported[29]. After extended exposures, film layers of goethite (α-Fe_2O_3,H_2O) and magnetite (Fe_3O_4) were formed. In marine environments these corrosion products are mixed with fouling deposits consisting of bacteria, microalgae and other higher organisms embedded in EPS. The cohesive effect of EPS in biofouling deposits depends on several environmental and biological factors and will influence the degree of biocorrosion/biofouling interactions. Observation of the microorganisms within the biofilm by electron microscopy is difficult to achieve on active metal surfaces and care must be taking when such surfaces are used in monitoring[30].

To illustrate the complex biocorrosion/biofouling interaction on carbon steel surfaces, laboratory experiments with a marine *Vibrio* and SRB in a saline medium will be described[31]. These experiments showed that, after a few hours of exposure, the metal samples were covered by a surface film of oxide above which a complex layer of amorphous corrosion products were formed. Isolated bacterial cells could be hardly seen in those conditions on the metal surface. After longer exposures, discrete colonies of bacteria were clearly seen on the surface. When these colonies were detached, inorganic crystals of haematite (Fe_2O_3) were found in the areas formerly covered by microbial colonies. If the removal procedure was continued until the complete removal of biological and inorganic deposits, an area of localised attack by micropitting was clearly distinguished were the bacterial colonies were located (Figure 5). When the exposure was extended to more than 3 days carbon steel specimens were densely covered by different

Figure 5 *SEM photograph showing a mild steel surface beneath a microbial colony of a marine Vibrio. after removing biological and inorganic deposits (bars indicate 10 mm).*

layers of corrosion products, impeding the visualisation of bacteria. Electrochemical experiments revealed favourable conditions for localised attack confirming the electron microscope observations. The preferential attack underneath microbial deposits was explained as the result of the following effects: i) differential aeration between metal areas covered and uncovered by the microorganisms and ii) depassivation of the metal surface through the reduction of insoluble ferric compounds into soluble ferrous ones as reported for certain types of bacteria[32]. The latter effect can be enhanced by the formation of microbial consortia within the biofilm structure, leading to an increase in the concentration and variety of metabolic corrosive species.

When compared with corrosion-resistant metal surfaces, biocorrosion/biofouling interactions reveal a higher complexity of effects and structures and give rise to more problems in interpretation using electron microscopy.

3. THE ENNOBLEMENT EFFECT IN STAINLESS STEELS

Although there are several interpretations for the ennoblement of the corrosion potential of stainless steel in natural seawater, the consensus of opinion is that this effect is due to a change in the cathodic reaction taking place on the metal and also to the active microbial participation in that change.

A complete review of the biological and electrochemical interpretations of this effect on stainless steel in seawater was recently published[22]. The influence of microorganisms on the corrosion of stainless steel can be twofold: i) an enhancement of the cathodic reaction with the consequent ennoblement of the corrosion potential, and ii) direct initiation of pitting under discrete biodeposits. The role of the primary biofilm can be confirmed by comparing the corrosion potential behaviour of stainless steel in natural seawater with the value in (0.2 mm pore diameter) filtered seawater where the majority of the film-forming bacteria have been removed. With this approach, it has been clearly shown by several authors[22,33] that the corrosion potentials of the control samples immersed in the filtered water did not show this effect. Other experimental approaches to decrease the ennoblement include increasing the flow speed of seawater[34] or raising the temperature sufficiently to inactivate the bacteria within the biofilm[35]. A different way to inhibit the effect of the biofilm on the ennoblement was made by adding a respiration inhibitor, sodium azide, to the system[33]. It has been demonstrated that the mere physical presence of the biofilm was not enough to cause ennoblement if an active metabolic process was not active in the system.

Sunlight and photosynthetic algae within the biofilm are detrimental to ennoblement. This effect is less likely to occur when light is present, or photosynthetic algae represent a significant portion of the biofilm[36]. These findings are important to the industrial uses of stainless alloys (storage tanks, pipes, heat exchangers, oil platforms) involve permanently immersed surfaces in dark water, where ennoblement is most readily observed. Conversely, corrosion measurements, either in the laboratory or the field, are frequently made in daylight, when the effect is reduced[22]. This would explain the absence of ennoblement seen by other authors[37,38]. A photoelectrochemical approach based on light effects on the oxide film at the metal surface has been recently reported[39].

The ennoblement effect, initially reported for seawater, has also been found in fresh and brackish waters[36], where the amount of ennoblement decreased nearly linearly with increasing salinity. Another factor influencing the ennoblement of the corrosion potential is linked to the degree of coverage of the metal surface by the biofilm. Whereas some authors claim that ennoblement in seawater is not achieved at less than 30% coverage[36], recent results[40] report ennoblement in fresh water for biofilm accumulation less than 60 mm in thickness and covering less than 20% of the exposed surface area.

The different mechanisms for interpreting ennoblement include: i) a catalytic enhancement of the oxygen reduction by organometallic complexes[41], ii) bacterially practised enzymes[42], iii) a modification of the oxygen reaction by a change of pH beneath the biofilm[36], iv) the induction of new cathodic reactions[42] or v) a modification of the surface film on the metal surface[39,40]. The main consequences for the corrosion behaviour of passive metals and alloys in seawater are: i) localised corrosion initiation and propagation, ii) an enhancement of the cathodic behaviour of stainless steel in galvanic corrosion for anodes such as copper alloys or iii) an increase in the current density involved with cathodic protection.

Although a definite mechanism for interpreting the ennoblement effect is still under discussion, the mechanism involving a combination of low pH, low dissolved oxygen levels and low (millimolar) concentrations of hydrogen peroxide at the metal/solution interface has proven to be effective for explaining the different experimental findings reported in the literature.

4. CATHODIC PROTECTION AND BIOFILM INTERACTIONS IN SEAWATER

Cathodic protection is usually accomplished by impressing an external current to the metal structure to be protected. In this manner, the applied current opposes the naturally occurring corrosion current. To achieve this situation, the metal structure should be polarised to a pre-selected potential in the immunity region at which it will be protected. Thus, the cost of this protective method will mainly depend on the amount of the current to be applied.

Cathodic protection alters the local chemistry at the metal surface inducing an increase in pH, due to the production of hydroxyl anions. This alkalinity reduces the solubility of calcium and magnesium compounds in the medium, favouring the precipitation of a calcareous scale[43]. In the presence of these deposits, a concentration polarisation is established and the current requirements to maintain the selected potential are reduced, decreasing the protection costs. The interactive effects between calcareous scale and biofilms, when applying cathodic protection to metal structures immersed in seawater, has received limited attention in the literature. The effect of the applied current on bacterial attachment and biofouling settlement is highly relevant to biocorrosion. According to recent publications[44,45], cathodic protection seems to be effective for controlling the growth of aerobic bacteria in carbon steel structures in seawater, whereas it will favour the growth of SRB in anaerobic biofilms. The content of organic material in seawater affects both the current requirements and the nature of the calcareous deposits formed at the metal/water interfaced. A pre-existing biofilm can render such deposits more uniform at all current densities. The combined bacterial/calcareous film acts as a beneficial diffusion barrier at high current densities but as a detrimental cathodic depolariser, increasing the current needed for protection, at low current densities. A primary marine biofilm can either enhance or decrease the effectiveness of cathodic protection, depending on the applied current density[46].

Cathodic protection reduces bacterial adhesion and reproduction during the early stages of biofouling settlement. When a steady state of biofilm growth has been reached, the effects induced by the cathodic current on bacterial biofilms are less relevant. Thus, at low temperatures, bacterial growth is lower and the effect of cathodic protection is more noticeable because the time to reach a steady state of biofilm settlement is longer[47,48].

The enhancement of anaerobic biofilms growth on cathodically protected steel surfaces may be attributed to enhanced hydrogen production at the metal surface under those circumstances. When a stable SRB biofilm is formed on stainless steel surfaces in seawater, cathodic protection seems to be unable to avoid localised corrosion initiation even at very negative potentials[48].

5. NEW STRATEGIES FOR MONITORING BIOCORROSION AND BIOFOULING ON OFFSHORE STRUCTURES AND COASTAL POWER PLANTS

One of the most significant problems encountered in offshore oil production, has been the occurrence of corrosion related to microbial growth and settlement on metal surfaces[49]. The frequency and importance of biocorrosion and biofouling effects on corrosion depend on: i) the velocity, temperature, pressure and oxygen content of the injection water and ii) the physico-chemical characteristics of the seawater, such as its organic content, oxygen, pH and chemical composition.

A monitoring programme should include parallel field and laboratory measurements, supported by appropriate sampling devices. Owing to the differences in biofouling and biocorrosion, an effective monitoring programme must necessarily provide information on water quality, corrosive attack, sessile and planktonic bacteria, biofilm characteristics, and the chemical composition of inorganic and biological deposits[50]. Monitoring devices should permit discrimination between inorganic fouling and biofilms, provide information on the nature and diversity of microbial components of biofouling and allow sampling of the deposits. On the other hand, sampling devices should fulfil several requirements: i) to ensure that metal coupons experience a flow regime similar to the rest of the pipe surface, ii) to be inexpensive and simple to manufacture, iii) to be easily withdrawn from the system, iv) to hold several sampling coupons for duplication of tests and a higher diversity of assays per sampling device, v) to be easily fitted in any conventional access to the system pipework or laboratory flow loops and vi) when used in pressurised lines to be designed for compatiblility with existing high-pressure fittings, to avoid partial shut-downs and depressurisation[30].

A monitoring programme, based on a variety of devices and analytical techniques (Table 1) and using different monitoring frequencies and equipment locations, was recently applied to assess biofouling settlement and biocorrosion in two different offshore oil production platforms in the South Atlantic Ocean[51].

Table 1 *Classification for monitoring field devices and analytical techniques in biocorrosion and biofouling*

System	Localisation	Field devices and analytical techniques	Experimental measurements
I	1	Coupons Racks 1020 steel coupons	Corrosion rates/mm y^{-1}
II	2		
I	A	Field test kits (in-situ chemical analysis)	Oxygen/ppb
II	B		Hydrogen sulphide/ppm
I	A	Water analysis	pH, pAlk, Total hardness,Ca, Mg, Cl^-, SiO_2, SO_4^{2-}, Total Fe, Total dissolved solids, Ba, Sr
II	B		
I	1	Biprobe (N80 and carbon steel coupons)	Scanning electron microscopy (SEM)
II	2		
II	-	Biofouling monitor (N80 and carbon steel coupons)	Most probable number of colony former units per unit area/CFU cm^{-2}
I	A	Microbial counts of planktonic SRB	Most probable number of colony former units per unit volume/ CFU cm^{-2}
II	B		

In this practical case it was possible to study how marine biofouling interacted with increasing amounts of corrosion products during the exposure of steel structures to seawater. Corrosion data from weight loss measurements and corrosion probes in the field were complemented with potentiodynamic runs and corrosion potential vs. time measurements in the laboratory. For biofouling assessment, two types of monitoring devices were used: an on-line bioprobe in the pressurised part of the line and a side-stream biofouling monitor (Figure 6). Scanning electron microscopy (SEM) observations and EDX analysis of the deposits were carried out on coupons removed from these systems.

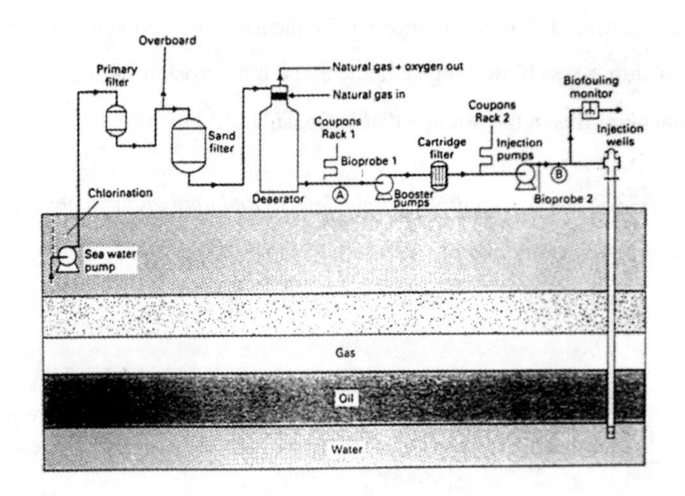

Figure 6 *Scheme of the offshore seawater injection system mentioned in the text showing the location of corrosion and biofouling monitoring devices, (After Videla et al.[51]) with permission of NACE International, Houston, TX.*

The results of this monitoring program, used to establish the working conditions of the system, prior to the development of a biocide strategy, show: i) a poor corrosion resistance for the structural steel of the system, ii) the high corrosion rates measured induced complex interactions between the corrosion products and biofilms, iii) SEM observations detected pitting and general corrosion in the metal samples where biofouling settlement was more intense and iv) in all cases the corrosive characteristics of seawater itself were inadequate to explain the degree and type of attack observed on the metal surfaces.

Coastal power plants cooling water systems are also prone to biocorrosion and biofouling. Most cases of biocorrosion in the power industry present similar characteristics, including the formation of discrete biological deposits in the internal wall of heat exchanger tubes, water conduits or storage tanks[2,52]. Pitting and crevice corrosion are generally found beneath these deposits. The deleterious effects of biofouling in heat transfer surfaces of the condenser are: i) an increase in frictional resistance to water flow; ii) a decrease of heat transfer efficiency and, iii) the corrosion of heat exchanger pipes. Macrofouling settlement on the precondenser cooling water structure produces flow reduction, tube blockage, increased microfouling, mechanical damage and erosion/corrosion effects. Biofouling and biocorrosion monitoring results have been reported for a coastal power plant heat exchanger system, located near the Mar del Plata Harbour in Argentina[52]. The seawater feeding this power plant was highly polluted due to a lack of water renewal and the waste discharges from the fishing industry into the harbour area. Thus, low dissolved oxygen levels and high sulphide contents were found. Biofouling components were mainly species frequently found in polluted waters[26]. Annual dominant species were the protozoan *Zoothamnium sp.* and the diatom *Navicula sp.* Subdominant species were: *Skeletonema costatum, Amphora exigua, Nematoda and Pinnulada sp.*

To study biocorrosion/biofouling interactions, stainless steel (AISI 304, 316 and 430) and 70:30 copper nickel samples were exposed for several weeks to flowing seawater (approximately 0.3 m s^{-1} at an average temperature of 15 °C) in the intake canal of the plant. The corrosion behaviour of the different metals was studied in the laboratory by using potentiodynamic polarisation and corrosion potential vs. time measurements. SEM observations were made to study the biofouling characteristics and/or metal attack after removal of deposits from the metal surface.

A summary of the main conclusions drawn from this practical case are: i) stalked ciliates, like *Zoothamnium sp.,* can facilitate the detachment of passive layers via adhesion effects developed at the fixation points of their peduncles assisted by water flow, ii) differential aeration was the predominant type of corrosion found in the areas of biofouling detachment for the copper-nickel alloy, iii) stainless steel samples were easily and rapidly colonized by microorganisms (micropitting, mainly associated with inclusion areas, was frequently found), iv) sulphide anions enhanced localised corrosion on copper nickel alloys and v) biodeterioration of all metals tested were directly related to the intensity of biofouling and corrosion products interactions at the metal/seawater interface.

6. NEW PERSPECTIVES IN THE STUDY OF BIOCORROSION IN THE MARINE ENVIRONMENT

A dramatic improvement in analytical, microbiological, electrochemical and microscopical instrumentation has occurred in the last decade, opening new perspectives for the study of biocorrosion and biofouling effects in the marine environment. In addition, methods for laboratory and field assessment of biocorrosion effects have been reported[19,53-55].

Chemical analysis and monitoring inside the biofilm thickness by means of specific microsensors is one of the most exciting new fields. These sensors complemented with the CSLM allow the assessment of the relationship between biofilm internal structure and oxygen profiles. A practical application of these techniques has been used to discuss the use and limitations of oxidising biocides on bacterial biofilms[56]. Results show that the chlorine concentration measured in the biofilm was typically $<20\%$ of the concentration in the bulk liquid. The shape of the chlorine profiles, the long equilibration times and the dependence on the bulk biocide concentration showed that chlorine penetration is a function of simultaneous reactions and diffusion within the biofilm matrix. Thus, the limited penetration of chlorine into the biofilm would be an important factor influencing the efficacy reduction observed for this biocide against biofilms, when compared with its action on planktonic cells. A similar reduced biocidal efficacy on sessile bacteria, recently reported for ozone[15], would be explained in the light of the new concepts on biofilm heterogeneity developed by the use of microsensors. In the case of ozone, its high oxidizing power would induce an alteration in the EPS and corrosion products in the outer layers of the biofilm, forming a barrier to further penetration of ozone to reach inner colonies (or clusters) of sessile bacteria by blocking the channels within the biofilm structure. This assumption is supported by CSLM data[57] showing how bacterial biofilms exposed to the antimicrobial agent fleroxacin suffer a disruption of biofilm channels by the antibacterial agent. Thus, the information obtained from the bulk water analysis is limited, and must be closely analysed before any final conclusions are drawn.

Advanced microbiological techniques, such as DNA probes, have been complemented with field data for monitoring biocorrosion in oil production through the assessment of the SRB community structures. Advanced microscopical instruments, like the CSLM and the Atomic Force Microscope (AFM), have allowed the examination of hydrated biofilms by means of non-intrusive techniques yielding clean three-dimensional images of living biofilms in real time[55,58]. In practice, innovative electrochemical methods for monitoring biocorrosion in industrial structures[59], as well as new field electrochemical devices to monitor biocidal application for preventing biocorrosion of steel in seawater, have been reported[60].

7. CONCLUDING REMARKS

A thorough understanding of the interactions between biofilms and corrosion products is essential for the interpretation of biocorrosion and biofouling problems in the marine environment. The corrosion behaviour of metal surfaces in seawater will vary according to the intensity and nature of this reciprocal interaction.

New insights into biofilm systems, developed by using innovative techniques and instrumentation, have changed classical concepts on biofilm structure and mass transport within the biofilm matrix.

New strategies for monitoring biocorrosion and biofouling in offshore structures and coastal installations should be addressed to elucidate the dynamic and complex interactions developed at biologically conditioned metal/solutions interfaces exposed to seawater.

Acknowledgements

Special thanks are due to Fundación Antorchas, The British Council and the Organising Committee of the 9th International Congress on Marine Corrosion and Fouling for their financial contributions.

References

1. W.G. Characklis and K.C. Marshall. Biofilms: a basis for an interdisciplinary approach, in: W.G.Characklis and K.C.Marshall (eds.), Biofilms, Wiley Interscience, New York., pp. 3-15, 1990.
2. R.G.J. Edyvean, and H.A. Videla. Biofouling and MIC interactions in the marine environment in: A.K.Tiller and C.A.C.Sequeira (eds.). Microbial Corrosion, EFC Publication No. 8, London, pp. 18-32, 1992.
3. H.A. Videla. Electrochemical aspects of biocorrosion, in C.C.Gaylarde and H.A.Videla (eds.), Bioextraction and Biodeterioration of Metals, Cambridge University Press, Cambridge, pp.85-127, 1995.
4. H.A. Videla and W.G. Characklis, Biofouling and microbiologically influenced corrosion. *Int. Biodet.Biodegr.*, 1992, **29,** 195.
5. H.A. Videla. Metal dissolution/redox in biofilms, in: W.G.Characklis and P.Wilderer (eds.), Structure and Function of Biofilms, John Wiley & Sons, Chichester, U.K. pp. 301-320, 1989.
6. S. Abkowitz, B. Bannon, R. Broadwell, C.E. Forney, Jr., B. Harvey, W. Herman, R. Kane, W. Minkler, J.R. Newman, J.R. Schley, R. Schultz, and K. Soltow. Why do we use titanium?, in: W.E.Herman, R.Broadwell, H.D.Kessler, J.Monses and G.M.Hockaday (eds), Titanium, the Choice, Titanium Development Association, Dayton, OH., pp. 2-8, 1990.
7. F.M. Reinhart. Corrosion of materials in hydrospace. Part III, Titanium and titanium alloys. U.S.Naval Civil Engineering Lab. Tech. Note N-921. Port Hueneme, CA, 1967.
8. H.A. Videla. Biocorrosion of nonferrous metal surfaces, in: G.G.Geesey, Z.Lewandowski and H.C.Flemming (eds.), Biofouling and Biocorrosion in Industrial Water Systems, Lewis Publishers, Boca Raton, FL, pp.231-241, 1994.
9. R.W. Schutz. A case for titanium's resistance to microbiologically influenced corrosion, *Mater.Perform.*, 1991, **30,** 58.
10. H.A. Videla, S.G. G6mez de Saravia and M.F.L. de Mele. MIC of heat exchanger materials in marine media contaminated with sulphate-reducing bacteria. Paper No. 189, Corrosion 92, NACE International, Houston, TX, 1992.

11. J.B. Cotton and B.P. Downing. Corrosion resistance of titanium to seawater. *Trans.Inst. Marine Engineering,* 1957, **69,** 311.
12. W.L. Adamson. Marine fouling of titanium heat exchangers, Report PAS-75-29, David W.Taylor Naval Ship Research and Development Center, Bethesda, MD, 1976.
13. B.J. Little, P.A. Wagner and R.I. Ray. An evaluation of titanium exposed to thermophilic and marine biofilms. Paper No. 308, Corrosion 93, NACE International, Houston, TX, 1993.
14. P.J. Aylott, J.F.D. Stott, R.D. Eden and H.K. Grover. Monitoring of marine biofouling of titanium tubed heat exchanger using a remote controlled thermal resistance method. Paper No. 195. Corrosion 95, NACE International, Houston, TX, 1995.
15. H.A. Videla, M.R. Viera, P.S. Guiamet, M.F.L. de Mele and J.C. Staibano Alais. Effect of dissolved ozone on the passive behaviour of heat exchanger structural materials. Biocidal efficacy on bacterial biofilms. Paper No.199. Corrosion 95, NACE International, Houston, TX, 1995.
16. H.A. Videla, M.F.L. de Mele and G.J. Brankevich. Microfouling of several metal surfaces in polluted seawater and its relation with corrosion. Paper No. 365, Corrosion 87, NACE International, Houston, TX, 1987.
17. Z. Lewandowski, W.C. Lee, W.G. Characklis and B.J. Little. Microbial alteration of the metal water interface: dissolved oxygen and pH microelectrode measurements. Paper No. 93, Corrosion 88, NACE International, Houston, TX, 1988.
18. Z. Lewandowski, F. Roe, T. Funk, and D. Chen. Chemistry near microbially colonised metal surfaces, in: H.A.Videla, Z.Lewandowski and R.W.Lutey, Biocorrosion & Biofouling : Metal/Microbe Interactions, Buckman Laboratories International, Memphis, TN, pp. 52-61, 1993.
19. Z. Lewandowski, P. Stoodley and F. Roe. Internal mass transport in heterogeneous biofilms. Recent advances. Paper No. 222, Corrosion 95, NACE International, Houston, TX, 1995.
20. S.C. Dexter, K.E. Lucas and G.Y. Gao. The role of marine bacteria in crevice corrosion initiation, in : S.C.Dexter (ed.), Biologically Induced Corrosion, NACE

International, Houston, TX. pp. 144-153, 1986.

21. A. Mollica. Biofilm and corrosion on active-passive alloys in seawater. *Int. Biodet. Biodegr.,* 1992, **29,** 213.
22. S.C. Dexter. Effects of biofilms on marine corrosion of passive alloys, in: C.C.Gaylarde and H.A.Videla (eds.), Bioextraction and Biodeterioration of Metals, Cambridge University Press, Cambridge, pp. 129-167, 1995.
23. G. Blunn. Biological fouling of copper and copper alloys, in: S.Barry, D.R.Houghton, G.C.Llewellyn and C.E. O'Rear (eds.), Biodeterioration 6, CAB International, pp. 567-575, 1986.
24. W.G. Characklis and K.E. Cooksey. Biofilms and microbial fouling. *Adv. Appl. Microbiol.,* 1983, **29,** 93.
25. H.A. Videla. Biocorrosion and biofouling. Metal/Microbe interactions. A retrospective overview. In: H.A.Videla, Z.Lewandowski and R.W.Lutey (eds.), Biocorrosion and Biofouling. Metal/Microbe Interactions, Buckman Laboratories International, Memphis, TN, pp. 101-108, 1993.
26. H.A. Videla, M.F.L. de Mele and G.J. Brankevich. Biofouling and corrosion of stainless steel and 70/30 copper-nickel samples after several weeks of immersion in seawater. Paper No. 291. Corrosion 89 NACE International, Houston, TX, 1989a.
27. M.F.L. de Mele, H.A. Videla and G.J. Brankevich. Corrosion of CuNi 30 Fe alloy in artificial solutions and natural seawater. Influence of biofouling. *Br.Corros.* J., 1987, **24,** 21 1.
28. A.H.L. Chamberlain and B.J. Garner. The influence of iron content on the biofouling resistance of 90/10 copper-nickel alloys. *Biofouling,* 1988, **1**, 79.
29. R.G.J. Edyvean. Interactions between microfouling and the calcareous deposit formed on cathodically protected steel in seawater, in: Proc.6th.Int. Congr. on Marine Corrosion and Fouling, Athens, pp. 469-483, 1984.
30. H.A. Videla, F. Bianchi, M.M.S. Freitas, C.G. Canales and J.F. Wilkes. Monitoring biocorrosion and biofilms in industrial waters: A practical approach, in: J.Kearns and B.J.Little (eds.), Microbiologically Influenced Corrosion Testing, ASTM STP 1232, Philadelphia, PA, pp. 128-137, 1994.
31. C.C. Gaylarde and H.A. Videla. Localised corrosion induced by a marine

Vibrio, Int. Biodet., 1987, **23,** 91.

32. W.C. Ghiorse. Microbial reduction of manganese and iron, in: A.J.B.Zehnder (ed.), Biology of Anaerobic Microorganisms, Wiley Interscience, New York, pp. 303-331, 1988.
33. V. Scotto, R. DiCintio and G. Marcenaro. The influence of marine aerobic microbial film on stainless steel corrosion behaviour. *Corros. Sci.*, 1985, **25,** 185.
34. A. Mollica and A. Trevis. Correlation entre la formation de la pellicule primaire et la modification de la reaction cathodique sur des aciers inoxydables : experimentes en eau de mer aux vitesses de 0.3 a 5.2 m/s. In: Proc. 4th Int.Congr. Marine Corrosion and Fouling, Juan-Ies-Pins, Antibes, France, pp. 351-365, 1976.
35. A. Mollica, A. Trevis, E. Traverso, G. Ventura, G. De Carolis and R. Dellepiane. Cathodic performance of stainless steel in natural seawater as a function of microorganisms settlement and temperature, *Corrosion,* 1989, **45,** 48.
36. S.C. Dexter and H.J. Zhang. Effects of biofilms on corrosion potential of stainless alloys in estuarine waters, Proc.11th Intern.Corr.Congr., Florence, Italy, pp. 4-333, 1990.
37. F. Mansfeld, H. Shih and C. Tsai. Results of exposure of stainless steel and titanium to natural seawater. Paper No. 109, Corrosion 90, NACE International, Houston, TX, 1990.
38. B.J. Little, R. Ray, P. Wagner, Z. Lewandowski, W. Lee, W.G. Characklis and F. Mansfeld. Impact of biofouling on the electrochemical behaviour of 304 stainless steel in natural seawater, *Biofouling,* 1991, **3,** 45.
39. S. Maruthamuthu, G. Rajagopal, S. Sathianarayannan, M. Eashwar and K. Balakrishnan. A photoelectrochemical approach to the ennoblement process: proposal of an adsorbed inhibitor theory. *Biofouling,* 1995, **8,** 223.
40. W. Dickinson and Z. Lewandowski. Electrochemical and microelectrode studies of stainless steel ennoblement. Paper No. 223, Corrosion 95, NACE International, Houston, TX, 1995.
41. R. Johnsen and E. Bardal. The effect of microbiological slime layer on stainless steel in natural seawater. Paper No. 227, Corrosion 87, NACE International, Houston, TX, 1985.

42. S.C. Dexter. Role of microfouling organisms in marine corrosion. *Biofouling,* 1993, **7**, 97.
43. G.Hernandez, W.H. Hartt and H.A. Videla. Marine biofilms and their influence on cathodic protection: a literature survey. *Corros. Reviews,* 1994, **12**, 29.
44. R.G.J. Edyvean. The effects of microbiologically generated hydrogen sulphide in marine corrosion. *MTS Joumal,* 1990, **24,** 5.
45. J. Guezennec. Influence of cathodic protection of mild steel on the growth of sulphate-reducing bacteria at 35 °C in marine sediments, *Biofouling,* 1991, **3,** 339.
46. S.C. Dexter and S.H. Lin. Calculation of seawater pH at polarised metal surfaces in the presence of surface films. *Corrosion,* 1992, **48,** 50.
47. H.A. Videla, S.G. Gómez de Saravia and M.F.L. de Mele. Early stages of bacterial biofilm and cathodic protection interactions in marine environments, in: Proc.12th Int.Corros.Congr. NACE International, Houston, TX, 5B pp. 3687-3695, 1993.
48. M.F.L. de Mele, S.G. Gómez de Saravia and H.A. Videla. An overview on biofilms and calcareous deposits interrelationships on cathodically protected steel surfaces. In: P.Angell, S.W.Borenstein, R.A.Buchanan, S.C.Dexter, N.J.E. Dowling, B.J.Little, C.D.Ludlin, M.B.McNeil, D.H.Pope, R.E.Tatnall, D.C.White and H.G. Ziegenfuss (eds.), Proceedings of 1995 International Conference on Microbially Influenced Corrosion, NACE International, Houston, TX, pp. 50/1-50/8, 1995.
49. R.G.J. Edyvean. Biodeterioration problems of North Sea oil and gas production. A review. *Int. Biodet.*, 1987, **23,** 199.
50. H.A. Videla, P.S. Guiamet, O.R. Pardini, E. Echarte, D. Trujillo and M.M.S. Freitas. Monitoring biofilms and MIC in an oilfield water injection system. Paper No. 103, Corrosion 91, NACE International, Houston, TX, 1991.
51. H.A. Videla, M.M.S. Freitas, M.R. Araujo and R.A. Silva. Corrosion and biofouling studies in Brazilian offshore seawater injection systems, Paper No. 191, Corrosion 89, NACE International, Houston, TX, 1989b.
52. G.J. Brankevich, M.F.L. de Mele and H.A. Videla. Biofouling and corrosion in coastal power plant cooling water systems, *MTS Joumal,* 1990, **24**, 18.
53. D.W.S. Westlake, G. Voordouw and T.R. Jack. Use of nucleic acid probes in

assessing the community structure of sulphate-reducing bacteria in Western Canadian oil field fluids, in: Proc.12th Int.Corr.Congr., NACE International, Houston, TX., **5B,** pp. 3794-3802, 1993.

54. G.G. Geesey, Z. Lewandowski and H.C. Flemming (eds.). Biofouling and Biocorrosion in Industrial Water Systems, Lewis Publishers, Boca Raton, FL, 1994.
55. A Steele, D.T. Goddard and l.B. Beech. An atomic force microscopy study of the biodeterioration of stainless steel in the presence of bacterial biofilms. *Int.Biodet. Biodegr.,* 1994, **34,** 35.
56. D. de Beer, R. Srinivasan and P.S. Stewart. Direct measurement of chlorine penetration into biofilms during disinfection. *Appl.Environ. Microbiol.,* 1994, **60,** 4339.
57. D.R. Korber, G.A. James and J.W. Costerton. Evaluation of Fleroxacin activity against established *Pseudomonas fluorescens* biofilms, *Appl.Environ.Microbiol.,* 1994, **60,** 1663.
58. J.W. Costerton. Structure of biofilms, in: G.G.Geesey, Z.Lewandowski and H.C.Flemming, Biofouling and Biocorrosion in Industrial Water Systems, Lewis Publishers, Boca Raton, FL, pp. 1- 14, 1994.
59. M.A. Winters, P.S.N. Stokes, P.O. Zuniga, and D.J. Schlottenmier. Developments in on-line corrosion and fouling monitoring in cooling water systems. Paper No. 392, Corrosion 93, NACE International, Houston, TX, 1993.
60. G. Ventura, E. Traverso and A. Mollica. Effect of NAClO biocide additions in natural seawater on stainless steel corrosion resistance, *Corrosion,* 1989, **45,** 319.

2. SURFACE ANALYTICAL TECHNIQUES APPLIED TO MICROBIOLOGICALLY INFLUENCED CORROSION

Brenda Little and Patricia Wagner

Naval Research Laboratory
Stennis Space Center, MS 39529-5004

Abstract

The study of microbiologically influenced corrosion has progressed from phenomenological case histories to a mature interdisciplinary science including electrochemical, metallurgical, microbiological, biotechnological, biophysical, and surface analytical techniques. Surface chemical techniques, including X-ray absorption and microelectrodes with tip diameters of a few microns, can determine, respectively, concentration and speciation of metal ions bound within exopolymers and can measure dissolved oxygen, pH, and other interfacial chemistries on the scale of microcolonies. Recent developments in image analysis systems-including electron, atomic force, and laser microscopy make it possible to image biological materials in the hydrated state and to accurately determine spatial relationships between microorganisms and localised corrosion. Scanning vibrating electrodes can be used to map the distribution of anodes and cathodes so that localised corrosion can be correlated with exact locations of microorganisms. Data derived from surface analytical techniques are used to define mechanisms of microbial corrosion.

1. INTRODUCTION

Corrosion associated with microorganisms has been recognized for over 50 years, yet the study of microbiologically influenced corrosion (MIC) is a relatively new, multidisciplinary field. Microorganisms growing on metal surfaces can produce crevices, differential aeration and metal concentration cells that result in pitting, and nonprotective corrosion products that can accelerate erosion corrosion and dealloying. MIC does not produce any unique type of corrosion that cannot be produced by an abiotic mechanism. Therefore, it is essential that a spatial relationship be established between causative microorganisms and localised corrosion.

Numerous mechanisms have been elucidated for the role of microorganisms in accelerating or controlling corrosion reactions. Microbiological and biochemical techniques have been traditionally used to determine physiologies of individual species from corroded surfaces. The problem has been that it is impossible to culture or detect all bacteria from a natural population, even with multiple media. The result is that the organisms that are easiest to grow have been most closely associated with MIC[1]. For example, sulphate reducing bacteria (SRB) capable of consuming cathodic hydrogen and concomitantly reducing sulphate to sulphide are ubiquitous and can be easily cultured from corroded metals exposed to seawater, fresh water, and soils. Microbiologically-produced sulphides react with metals to form non-protective corrosion products that are indicative of MIC[2]. However, there are numerous other microorganisms that cause corrosion independently or in consortia that cannot be cultured as easily. The ways in which surface analytical techniques are being used to understand mechanisms for MIC will be discussed in this chapter.

1.1 Surface Chemistry

It is now generally recognized that biofilms develop on all surfaces exposed to biologically active environments to produce biofilm/metal interfacial chemistries that can accelerate corrosion (Figure 1). Interfacial acids, bases and sulphides cannot be detected in the bulk medium. Ion-selective and gas-sensing microelectrodes with tip diameters less than 10 μm are being used for direct measurements within biofilms and at biofilm/metal interfaces. Lewandowski *et al.*[3] measured dissolved oxygen profiles in a continuous flow, open channel reactor with a mixed biofilm of a metal surface. They demonstrated that in an aerobic medium the biofilm/metal interface was anaerobic after the biofilm reached a thickness of 900 μm. That means that the rate of bacterial respiration is faster than the rate of oxygen diffusion through the same thickness.

Van Houdt *et al.*[4] developed a rugged iridium oxide pH microelectrode with a tip diameter in the range of 3 to 15 μm and measured pH profiles across a mixed population biofilm on a polycarbonate disc. Bacteria within biofilms can produce organic and inorganic acids that accelerate corrosion[5]. Isolated microbial colonies can also initiate under-deposit corrosion by creating a differential aeration cell that results in an acidic environment immediately under the respiring colony that is independent of microbial activity and dependent on metallurgy.

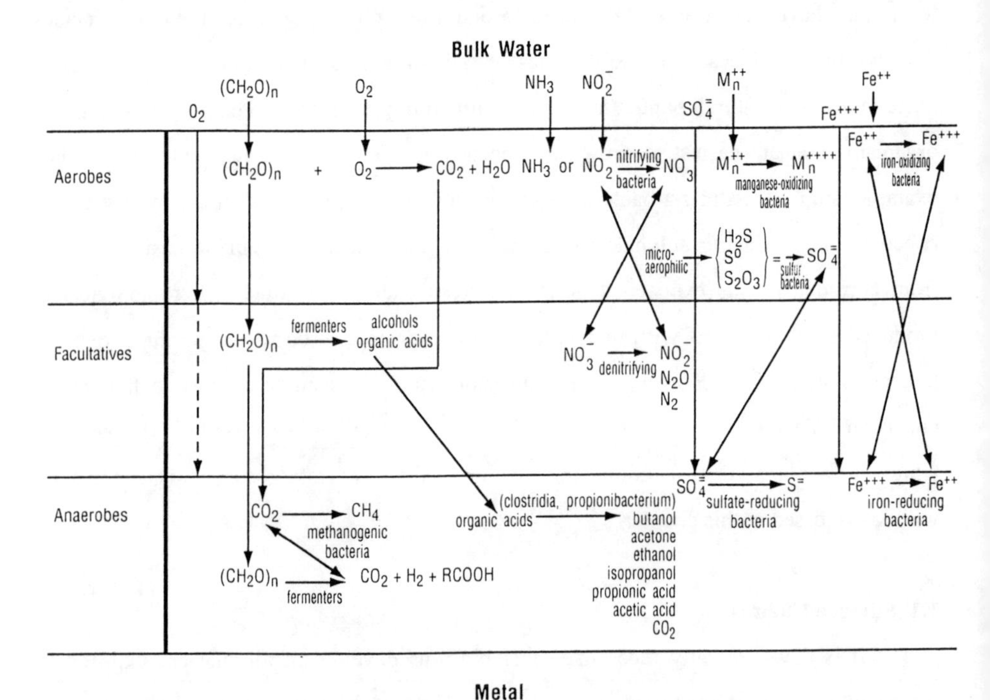

Figure 1 *Strata within a typical biofilm and possible reactions within the strata*

Briefly, in an oxygenated medium the area immediately under a respiring colony will become anodic and the anodic reaction will be metal dissolution (Figure 2). Adjacent areas will become cathodic and the cathodic reaction will be the reduction of oxygen. If the anode and cathode are separated, the pH at the anode will decrease and that at the cathode will increase. The exact pH at the anode is determined by the alloying elements

(Table 1). Lewandowski *et al.*[6] demonstrated pH decrease under an abiotic polymer on a carbon steel surface using the iridium oxide pH microelectrode (Figure 3).

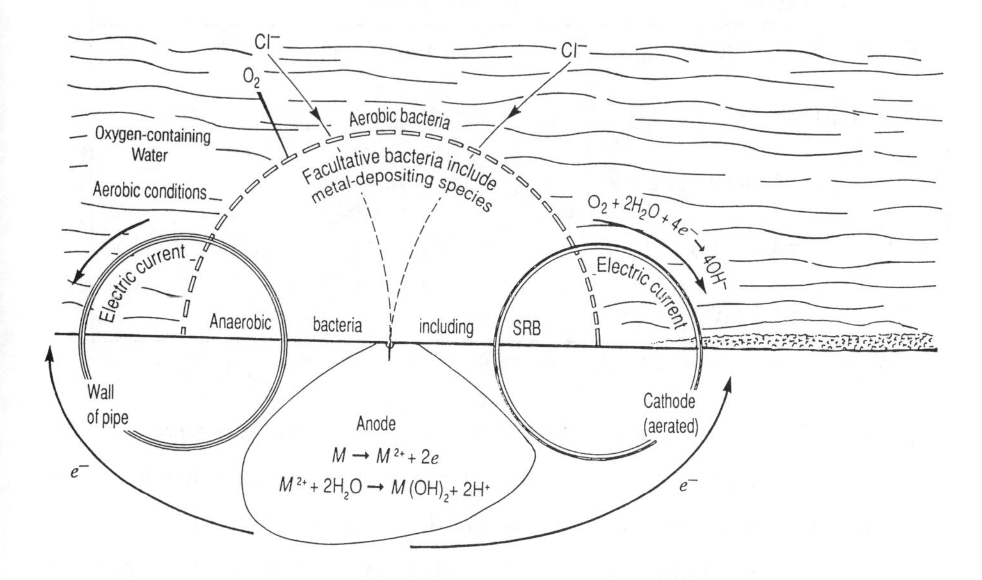

Figure 2 *Reactions under tubercles created by metal-depositing bacteria.*

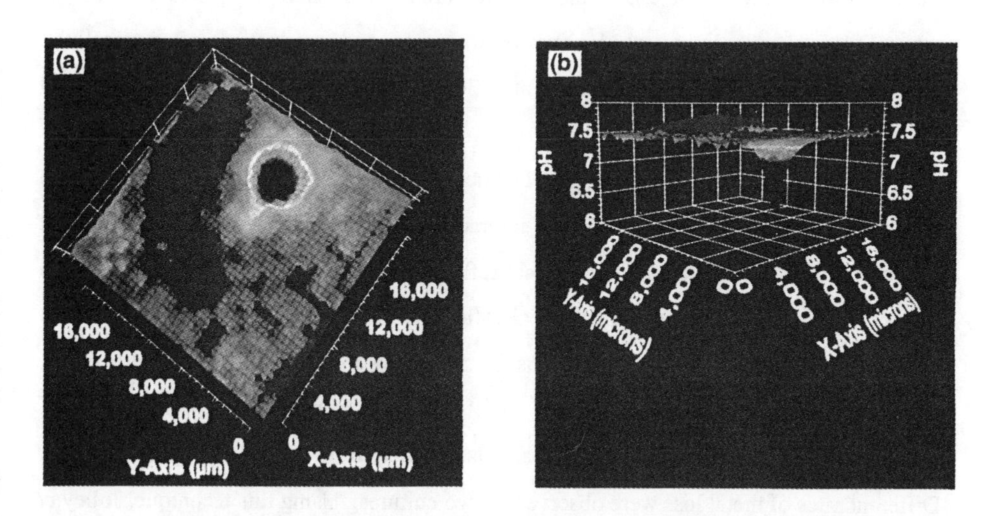

Figure 3 *Spatial distribution of pH on a mild steel surface using an iridium oxide microelectrode*[6]*, (a) planar pH distribution on surface partially covered with agar and (b)vertical pH profile of (a). ©ASTM. Reprinted with permission.*

Table 1 *Specific hydrolysis reactions*

Hydrolysis Reaction	Equilibrium pH
$Fe^{2+} + 2H_2O \rightleftharpoons Fe(OH)_2 + 2H^+$	$pH = 6{\cdot}64 - 1/2 \log a_{Fe^{2+}}$
$Cr^{3+} + 3H_2O \rightleftharpoons Cr(OH)_3 + 3H^+$	$pH = 1{\cdot}53 - 1/3 \log a_{Cr^{3+}}$
$Ni^{2+} + 2H_2O \rightleftharpoons Ni(OH)_2 + 2H^+$	$pH = 6{\cdot}5 - 1/2 \log a_{Ni^{2+}}$
$Mo^{3+} + 2H_2O \rightleftharpoons MoO_2 + 4H^+ + e^-$	$pH^* = (0{\cdot}311 - 0{\cdot}059 \log a_{Mo^{3+}} - E)/0{\cdot}236$
$Mn^{2+} + 2H_2O \rightleftharpoons Mn(OH)_2 + 2H^+$	$pH = 1{\cdot}53 - 1/3 \log a_{Mn^{2+}}$

$^*E = -0{\cdot}20$ V (*v.* SCE)

Nivens *et al.*[7] demonstrated that attenuated total reflectance infrared spectroscopy (ATR-FTIR) can be used to detect changes in sessile microbial biomass. They demonstrated that the number of attached *Caulobacter sp.* was directly correlated with the intensity of the infrared amide II asymmetrical stretch band at 1543 crn^{-1}, corresponding to bacterial protein. The technique was sensitive to 10^6 cm^2 bacteria, and changes in the physiological status of the attached bacteria could be measured. For example, production of the intracellular storage lipid poly-B hydroxyalkanoate and production of extracellular polymer were monitored by absorbance at 1730 cm^{-1} (C=O stretch) and 1084 cm^{-1} (C-O stretch), respectively.

Geesey and Bremer[8] used ATR-FTIR not only to detect biofilm formation, but also to nondestructively evaluate in real time interactions of bacteria isolated on thin films of copper evaporated onto germanium internal reflection elements. By sputter coating a thin film of copper on the germanium internal reflectance element, they were able to detect changes in the thickness of copper films by observing the increase in intensity of the infrared water absorption band at 1640 cm^{-1}. The authors compared copper loss from copper thin films in the presence of bacteria isolated from corroded copper samples. Different rates of metal loss were observed in two cultures. Using this technique, Jolley *et al.* [9] observed copper loss in the absence of bacteria due to metal binding by bacterial extracellular polymers. Differences of the order of 0.3 to 0.4 nm, the equivalent of two to three atomic layers of copper, could be detected.

Nivens *et al.* [7] investigated use of the quartz crystal microbalance (QCM), a very sensitive mass-sensing device, for detecting attached microbial films. The QCM was more sensitive to changes in biomass than ATR-FTIR, with a detection limit of 10^4 cm^{-2} bacteria and a linear range of at least two orders of magnitude. An interesting aspect of both ATR-FTIR and the QCM is that substrata of both techniques can be converted to electrodes for electrochemical analyses so that corrosion information can be obtained while changes in microbial biofilms are monitored.

Chamberlain *et al.* [10] and Geesey *et al.* [11] stressed the relationship between microbial exopolymers and copper corrosion in fresh water systems.

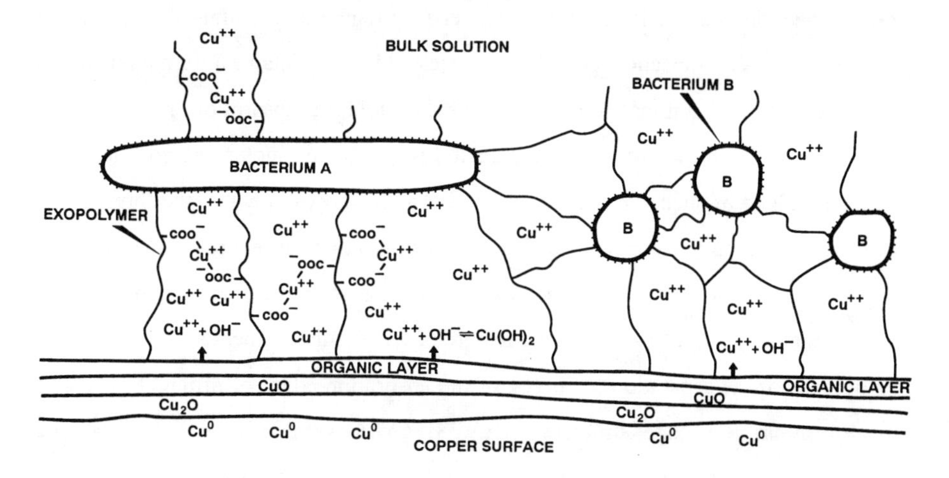

Figure 4 *Schematic representation of proposed copper concentration cells established on a copper surface coated with a mixed flora biofllm* [11] *©NACE International. All rights reserved by NACE; reprinted with permission.*

Each group emphasized the role of exopolymers in binding copper and the formation of copper concentration cells (Figure 4). Surface analytical tools have been used to resolve questions related to bound metals within biofilms. For example, energy dispersive spectroscopy (EDS) is an excellent tool for demonstrating the presence of metal ions within biofilms or dealloying as a result of biofilms, but cannot be used to quantify or determine the speciation of metal ions. Several investigators are attempting to determine the speciation of bound metals within cultures using X-ray photoelectron spectroscopy

(XPS) [12]. However, XPS cannot be used to evaluate metals bound within biofilms.

Wagner *et al.* [13] investigated the importance of copper concentration cells in marine corrosion using the bacterium *Oceanospirillum* which produces copious amounts of exopolymer when grown on copper surfaces and binds Cu^{2+} from the substratum. Little *et al.* [14], used X-ray absorption (XAS) to determine the speciation and concentration of copper within *Oceanospirillum* exopolymers. XAS is a technique for the investigation of electronic structure and local environment of specific atoms in liquids, solids, gases, solutions, and gels. An abrupt increase in absorption is observed when the X-ray energy is sufficient to liberate inner shell electrons. The absorption edge occurs at energies of several kiloelectronvolts for ls electrons. The energy region surrounding the absorption edge is less than the ionisation potential or threshold and contains information about charge density of the absorbing atom. Above the threshold, the absorption spectrum of an isolated atom gradually decreases monotonically as the X-ray photon energy increases. For atoms involved in chemical bonding, the absorption spectrum above the threshold is characterised by structure called extended X-ray absorption fine structure (EXAFS). XAS depends on precise measurement of the absorption cross section in the neighborhood of characteristic absorption edges. A unique feature of XAS is the element specificity which occurs because of the separation in energy of the absorption edges of different elements. X-ray absorption near edge structure (XANES) provides information on metal site symmetry, oxidation state and the nature of the surroundings; and EXAFS provides detail about type, number, and distances of atoms in the vicinity of the absorber. Little *et al.* [14] demonstrated that XANES spectra for Cu^{0}, Cu^{1+} and Cu^{2+}, were unique and that once Cu^{1+} was bound within a biofilm, further oxidation to Cu^{2+} did not take place. The role of microbiologically produced metal concentration cells in marine copper corrosion is still under investigation.

1.2 Microscopy

Recent developments in image analysis systems, including electron, atomic and laser microscopy, make it possible to image biological materials in the hydrated state without the usual dehydration steps required for SEM sample preparation. Using new microscopic techniques, studies by Rogers *et al.* [15] and Lewandowski *et al.* [16] provided the first new models of biofilm formation and architecture in the past two decades. Both demonstrated the heterogeneity and anisotropic nature of biofilms made up of distinct three dimensional microcolonies with water channels (Figure 5).

Little *et al.* [17] used environmental scanning electron microscopy (ESEM) coupled with EDS to study SRB associated with corroding copper alloys in seawater. The authors demonstrated that sulphides produced by SRB preferentially reacted with iron and nickel in the copper alloys, resulting in selective loss of these constituents and any corrosion inhibition they may have provided. SRB were encrusted with sulphides and were distributed throughout the sulphide corrosion layers (Figure 6). The tenacity of copper and nickel sulphide layers varied with alloy composition but all were easily removed with turbulence.

Iron-related bacteria (IRB) on stainless steel surfaces in fresh water are a much bigger problem than SRB. IRB produce dense tubercles of ferric hydroxide that create differential aeration cells and initiate under-deposit corrosion, previously described. To diagnose IRB as the cause of the under-deposit corrosion it is essential to establish their presence within the tubercles.

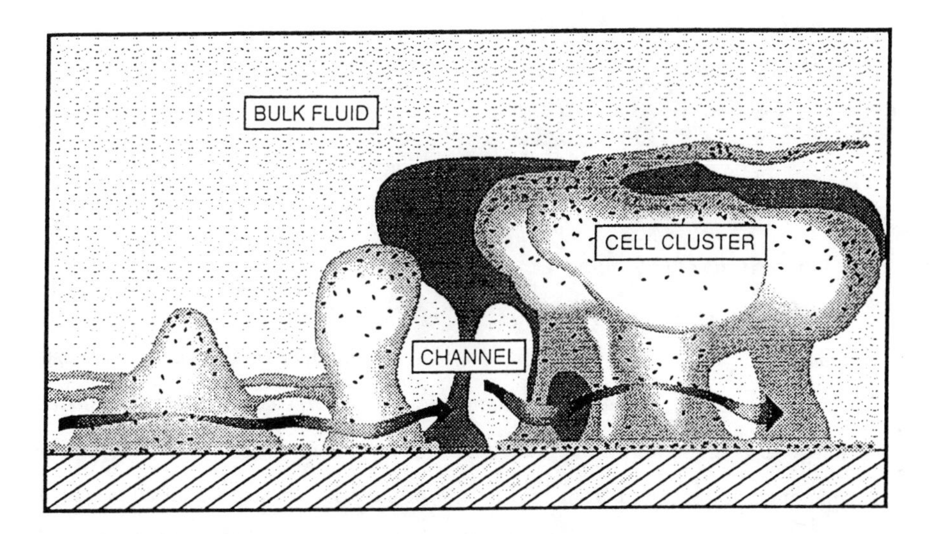

Figure 5 *Conceptual model of a biofilm, sagittal view, based on CLSM images [16]. ©NACE International. All rights reserved by NACE; reprinted with permission.*

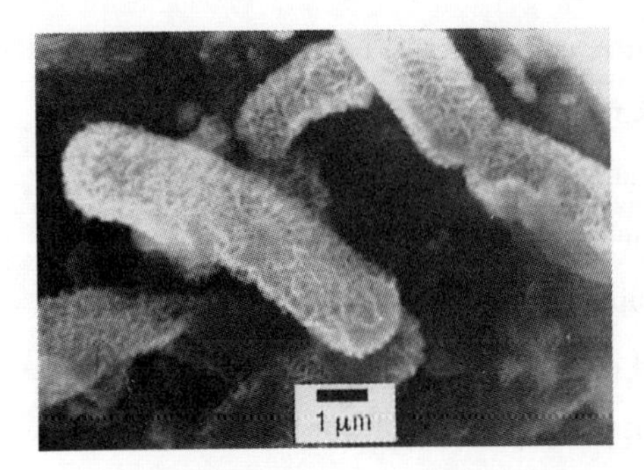

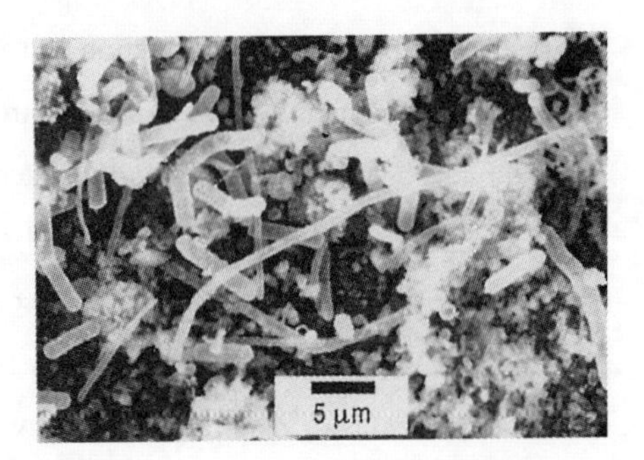

Figure 6 *(a) ESEM image of bacteria encrusted with copper sulphide crystals and (b) bacteria in copper sulphide corrosion layers*

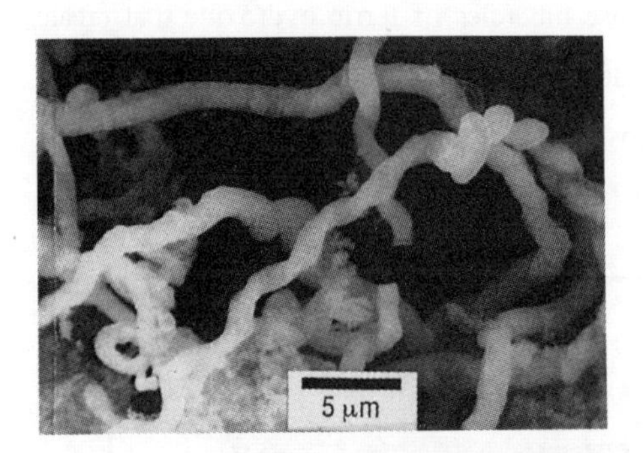

Figure 7 *ESEM image of intertwined microfibrils of IRB within a tubercle on a galvanized pipe surface*

IRB are difficult to isolate and identify using microbiological or biochemical procedures. The ESEM has been used to image twisted filaments of IRB within tubercles without any fixation (Figure 7). Some IRB are distinctive because of "characteristic elongated helical stalks composed of numerous intertwined microfibrils[18].

Confocal laser scanning microscopy (CLSM) permits one to create three dimensional images, see surface contour in minute detail and accurately measure critical dimensions by mechanically scanning the object with laser light[19]. A sharply focused image of a single horizontal plane within a specimen is formed while light from out-of-focus areas is repressed from view. The process is repeated again and again at precise intervals on horizontal planes and all images are compiled to create a single multidimensional view of the subject. Geesey used CLSM to produce three dimensional images of bacteria colonising copper and stainless steel surfaces. Colonisation was not uniform or random, but clearly associated with surface features, including scratches and milling lines[20] (Figure 8) and grain boundaries[21] (Figure 9).

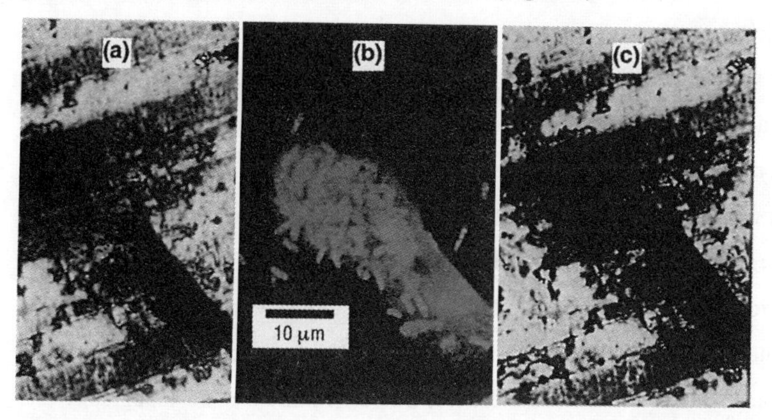

Figure 8 *CLSM image of a copper surface after exposure to a fresh water bacterium*[20] *(a) Reflected visible light image of surface showing milling lines (dark depressed areas), (b) reflected fluorescent light image of same area showing acridine orange-stained cells on surface, and (c) fluorescent image of (b)superimposed on (a) showing localization of cells in milling lines*

Reprinted with permission from G. G. Geesey and the American Society for Microbiology, Washington DC

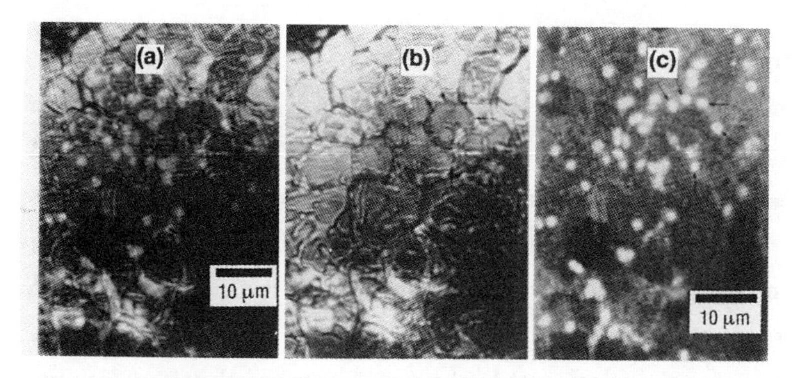

Figure 9 *CLSM images of 316 stainless steel exposed to Pseudomonas aeruginosa*[21] *(a) Reflected UV light image of surface containing attached acridine orange-stained cells, (b) reflected white light image of showing exposed grain boundaries, and (c) image (a) superimposed on (b) showing majority of cells to be associated with grain boundaries.*

Reprinted with permission from G. G. Geesey, 'Biofouling and Biocorrosion in Industrial Water Systems'(cover), G. G. Geesey, S. Lewandowski, and H-C.Flemming, eds. ©Lewis Publishers, an imprint of CRC Press, Boca Raton, FL, 1994

The atomic force microscope (AFM), derived from the scanning tunnelling microscope which uses an atomically sharp tip held angstroms from the surface to profile surface features with angstrom resolution. When the tip is electrically biased with respect to the sample, a current will flow between the surface atom closest to the tip and the nearest tip pattern by the quantum mechanical process of electron tunnelling. Current is exponentially sensitive to the distance between the tip and the sample. A piezoelectric transducer is used to mechanically scan the tip over the sample. AFM measures surfaces forces and provides exceptional topographical detail and allows viewing of specimens in the hydrated state. AFM uses a scanning probe to record x,y,z coordinates of a sample in fractions of a nanometer. Photodiode electrical outputs mimic sample topography and serve as the basis for the resulting image. AFM images of copper exposed to bacterial culture medium for 7 days showed biofilms distributed heterogeneously across the surface with regard to both cell numbers and depth[22]. Bacterial cells were associated with pits on the surface of the copper coupons. AFM was used by Steele *et al.*[23] to conclude that 316 stainless steel exposed to bacterial consortia showed a significant increase in surface roughness compared to uninoculated controls (Figure 10).

1.3 Electrochemistry

The scanning vibrating electrode technique (SVET) uses a vibrating electrode to map potential fields in solution over anodic and cathodic sites. Corrosion currents depend on polarisation characteristics of metal surfaces. The vibrating electrode converts potential fields into an alternating signal[24]. The SVET system consists of a 20 μm stainless steel microprobe electroplated with platinum black that is vibrated in two orthogonal directions while scanned under computer control over a coupon surface. The probe is capable of capacitively measuring local current density at each point of the scan, allowing the computer to generate maps of the local current densities over a surface. The vibrating probe system is mounted under a microscope fitted with reflected light, Nomarski and epifluorescence systems (40 x water objective, 2 mm working distance), and an imaging system. Current density scans can be concurrently mapped with images of the surface.

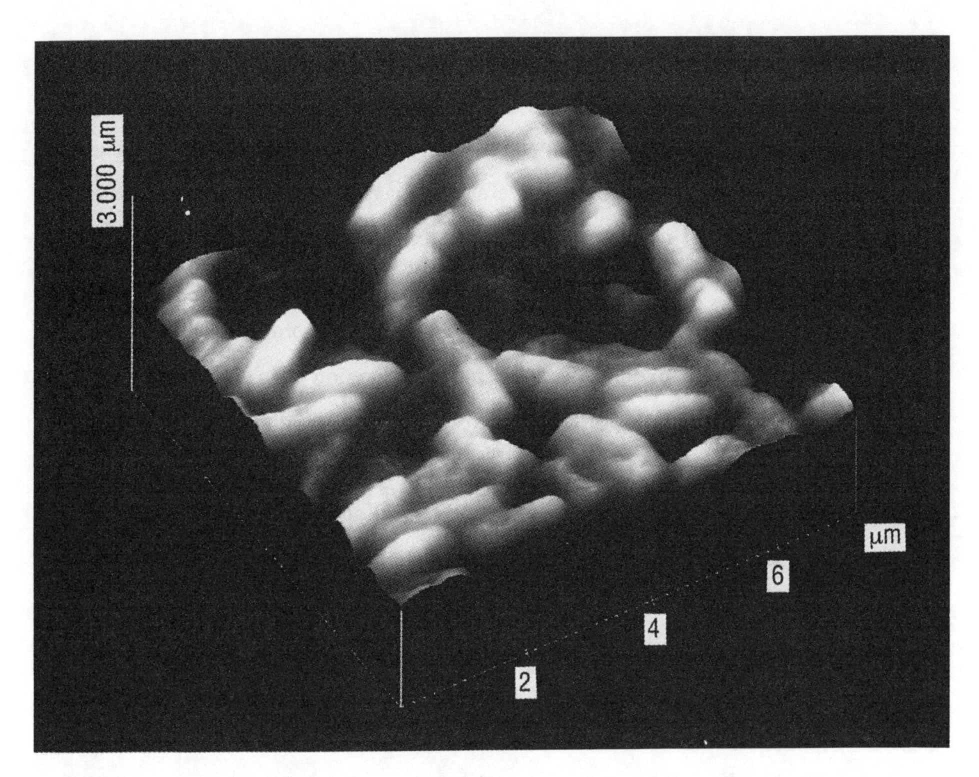

Figure 10 *AFM image of stainless steel exposed to a bacterial consortium.21 Provided courtesy of D. T Goddard, British Nuclear Fuelsplc, Preston, UK and A. Steele and l Beech, Chemistry Department., University of Portsmouth, UK*

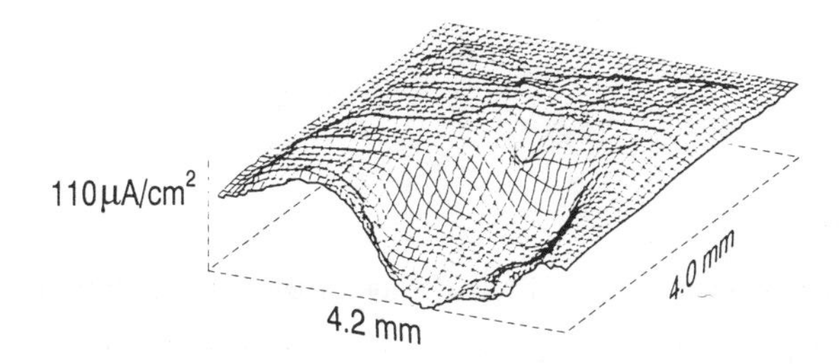

Figure 11 *SVET current density map over carbon steel after exposure to aerobic bacteria[25]. ©NACE International. All rights reserved by NACE; reprinted with permission.*

Franklin *et al.*[25] used SVET to demonstrate a spatial relationship between localised corrosion and bacterial cells on carbon steel (Figure 11). They demonstrated that in sterile, continuously stirred media, small pits initiated on steel surfaces subsequently repassivated. Initiated pits remained active and propagated in the presence of biofilms. Pit propagation depended on the presence of bacteria. The authors were careful not to conclude that bacterial cells caused pits. Instead, bacterial cells prevented repassivation and were clearly spatially related to the pits. The authors did propose that biofilms may inhibit migration of aggressive ions from the pits or migration of inhibiting ions from the bulk solution into pits. The sensitivity and resolution of SVET are determined by the distance between the probe and the metal surface, in addition to the solution conductivity. A decrease in the tip diameter increases the sensitivity and resolution of the technique. However, a very fine tip has a high impedance that increases electrochemical noise and decreases response time. Resolution also depends on the proximity of two corroding sites and the magnitude of the corrosion current from each site.

2. CONCLUSIONS

Surface analytical techniques provide the technologies necessary to discriminate microbially influenced corrosion by determining chemistries produced by microorganisms, as well as spatial relationships between microorganisms, corrosion products and corrosion currents.

Acknowledgments

This work was supported by the Office of Naval Research, Program Element 0601153N, through the Defense Research Sciences Program. Approved for public release, unlimited distribution, NRL Contribution Number NRL/JA/7333--95-0069.

References

1. B. Little and P. Wagner, 'Standard Practices in the United States for Quantifying and Qualifying Sulphate Reducing Bacteria in Microbiologically Influenced Corrosion', Proceedings Redefining International Standards & Practices for the Oil & Gas Industry, London, UK, 1992.
2. M. B. McNeil, J. M. Jones, and B.J. Little, 'Mineralogical Fingerprints for Corrosion Processes Induced by Sulphate Reducing Bacteria', Corrosion/91, paper no. 580, NACE International, Houston, TX, 1991.
3. Z. Lewandowski, S.A. Altobelli, and E. Fukushima, *Biotechnol. Prog.*, 1993, **9(1),** 40.
4. P. Van Houdt, Z. Lewandowski, and B. Little, *Biotechnol. Bioeng.*, 1992, **40,** 601.
5. J.M. Burns, E.E. Staffeldt, and O.H. Calderon, *Dev.Ind.Microbiol.*, 1969, **8,** 327.
6. Z. Lewandowski, T. Funk, F. Roe, and B. Little, in: 'Microbiologically Influenced Corrosion Testing', STP 1232, J. Kearns and B. Little, eds., ASTM, Philadelphia, PA, 1994, p. 61.
7. D.E. Nivens, J.Q. Chambers, and D.C. White, in: 'Microbially Influenced Corrosion and Biodeterioration', N.J. Dowling, M.W. Mittelman, and J.C. Danko, eds., University of Tennessee, Knoxville, TN, 1991, pp. 5-47.
8. G.G. Geesey and P.J. Bremer, *J. Mar.Tech.*, 1990, **24,** 36.
9. J.G. Jolley, G.G. Geesey, M.R. Hankins, R.B. Wright, and P.L. Wichlacz, *J. Appl. Spectros.*, 1989, 43, 1062.
10. A.H.L. Chamberlain, P. Angell, and H.S. Campbell, *Brit. Corr J.*, 1988, **23,** 197.
11. G.G. Geesey, M.W. Mittelman, T.Iwaoka, and P.R. Griffiths, *Mat.Perform., 1986,* **25,** 37.
12. J.R. Kearns, C.R. Clayton, G.P. Halada, J.B. Gillow, and A.J. Francis, 'The Application of XPS to the Study of MIC', Corrosion/92, paper no. 178, NACE International, Houston, TX, 1992.
13. P.A. Wagner, B.J. Little, and A.V. Stiffey, 'An Electrochemical Evaluation of Copper Colonised by a Copper-Tolerant Marine Bacterium', Corrosion/91, paper no.109. NACE International, Houston, TX, 1991.

14. B.J. Little, P.A. Wagner, K.R. Hart, R.I. Ray, D.M. Lavoie, W.E. O'Grady, and P.P. Trzaskoma, 'Localization and Speciation of Copper Ions in Biofilms on Corroding Copper Surfaces', Corrosion/94, paper no. 256, NACE International, Houston, TX, 1994.
15. J. Rogers, J.V. Lee, J.P. Dennis, and C.W. Keevil, 'Continuous Culture Biofilm Model for the Survival and Growth of *Legionella Pneumophila* and Associated Protozoa in Potable Water Systems', Proc.U.K. Symposium on Health-Related Water Microbiology, R. Morris, L. M. Alexander, P. Wyn-Jones, and J. Sellwood, eds., University of Strathclyde, Glasgow, 1991.
16. Z. Lewandowski, P. Stoodley, and F. Roe, 'Internal Mass Transport in Heterogeneous Bioflllms-Recent Advances', Corrosion/95, paper no. 222, NACE International, Houston, TX, 1995.
17. B. Little, P. Wagner, and J. Jones-Meehan, *Biofouling,* 1993, **6,** 279.
18. H.F. Ridgway, E.G. Means, and B.H. Olson, *Appl.Environ.Microbiol., 1981,* **41(1),** 288.
19. F.B. Baak, J.M. Thunnissen, C.B.M. Oudejans, and N.W. Schipper, *Appl.Optics,* 1987, **26,** 3413.
20. G.G. Geesey, ASM News (cover), **58(10),** 1992.
21. G.G. Geesey, 'Biofouling and Biocorrosion in Industrial Water Systems' (cover), G.G. Geesey, S. Lewandowski, and H.-C. Flemming, eds., Lewis Publishers, Boca Raton, FL, 1994.
22. P.J. Bremer, G.G. Geesey, and B. Drake, *Curr.Microbiol.,* 1992, **24,** 223.
23. A. Steele, D. Goddard, and 1. B. Beech, 'A Quantitative Study of the Corrosion Damage of Stainless Steel in the Presence of Bacterial Biofilms Using Atomic Force Microscopy', Proc.International Conference on Microbially Influenced Corrosion, P. Angell *et al.*, eds., NACE International, Houston, TX, 1995.
24. H. S. Isaacs and B. Vyas, 'Scanning Reference Electrode Techniques in Localised Corrosion', Pub. 727, ASTM, Philadelphia, PA, 1981.
25. M.J. Franklin, D.C. White, and H.S. Isaacs, 'The Use of Current Density Mapping in the Study of Microbial Influenced Corrosion', Corrosion/90, paper no. 104, NACE International, Houston, TX, 1990.

3. ELECTROCHEMICAL AND ATOMIC FORCE MICROSCOPY STUDIES OF A COPPER NICKEL ALLOY IN SULPHIDE-CONTAMINATED SODIUM CHLORIDE SOLUTIONS

G.J.W. Radford[a], F.C. Walsh[a], J. R Smith[a], C.D.S. Tuck[b] and S.A. Campbell[a],

[a]*Applied Electrochemistry Group, University of Portsmouth, St. Michael's Building, White Swan Road, Portsmouth, PO1 2DT (UK).* [b]*Meighs Ltd., Langley Alloys Division, Campbell Road, Stoke-on-Trent, Staffs, ST4 4ER (UK).*

Abstract

The electrochemical behaviour of copper and a high strength 70/20 copper nickel alloy, MARINEL™ (MARINEL™ is a registered trade make of Langley Alloys Ltd.), was investigated in sulphide contaminated and uncontaminated sodium chloride solutions (3.0%) at 293 K. Cathodic polarisation curves for the oxygen reduction reaction exhibited a well-defined plateau region between -650 and -1150 mV (*vs*.SCE) prior to hydrogen evolution. In the presence of 5 ppm sulphide ions, an additional plateau region between -480 and -850 mV (*vs.* SCE) is observed which is attributable to the production of hydrogen peroxide. This plateau is more prominent on copper than on the high strength alloy. The diffusion coefficient for dissolved oxygen was determined in the presence and absence of sulphide. Oxygen reduction was similar on copper nickel and copper leading to diffusion coefficients of 1.50×10^{-5} and 1.44×10^{-5} $cm^2\ s^{-1}$ respectively. In the presence of 5 ppm sodium sulphide, the diffusion coefficient values were reduced to 1.35×10^{-5} and 1.24×10^{-5} $cm^2\ s^{-1}$, respectively. Atomic force microscopy investigations of the initiation of corrosion and the influence of passive layers on the corrosion reaction confirmed that film formation was rapid. In the presence of sulphide ions, particle size increases from 68 to 300 nm in diameter were observed for both surface films but the particle heights were of similar dimensions, 6 nm. The increase in particle size for films produced in sulphide contaminated solutions would be expected to influence film porosity and hence the corrosion resistance.

Symbols	**Units**
I_L = Limiting current	A
n = Number of electrons involved in the electrode process	
F = Faraday constant	A s mol^{-1}
ω = Rotation rate of the disc	rad s^{-1}
ν = Kinematic viscosity of the electrolyte	cm^2 s^{-1}
c_B = Bulk concentration of dissolved oxygen	mol cm^{-3}
D = Diffusion coefficient of dissolved oxygen	cm^2 s^{-1}
A = Area of electrode	cm^2

1. INTRODUCTION

Copper and its alloys are used extensively as marine and engineering structural materials due to their excellent corrosion resistance and anti-fouling properties which result from their naturally occurring, protective corrosion products. Over the last 20 years, there have been rapid developments in their use as coatings and fasteners. Copper alloys, particularly cupro-nickels, are used as cladding materials in antifouling applications and the high strength, age hardening copper nickels have been widely used for offshore bolting applications[1].

It is generally accepted that the open circuit corrosion of freshly polished copper alloys involves the reduction of oxygen and the production of cuprous ions[2],

$$O_{2(g)} + 2H_2O_{(l)} + 4e^- \rightarrow 4OH^-_{(aq)} \quad \text{(cathodic reaction)} \tag{1}$$

$$Cu_{(s)} \rightarrow Cu^+_{(aq)} + e^- \quad \text{(anodic reaction)} \tag{2}$$

followed by the formation of a copper(I)-chloro complex,

$$Cu^{+}{}_{(aq)} + 2Cl^{-}{}_{(aq)} \rightarrow CuCl_2^{-}{}_{(aq)} \quad (3)$$

Cuprous oxide is widely acknowledged to be the initial product of corrosion[3-6],

$$2CuCl_2^{-}{}_{(aq)} + 2OH^{-}{}_{(aq)} \rightarrow Cu_2O_{(s)} + H_2O_{(l)} + 4Cl^{-}{}_{(aq)} \quad (4)$$

and is generally doped with alloying elements which can render it more protective than copper[7]. The presence of nickel and iron as minor components in the lattice of the corrosion layer on copper nickels decreases its defect number, thus increasing its passivating power[4].

After prolonged exposure to chloride environments such as seawater, further oxidation processes occur, leading to the formation of copper (II) compounds[8]. For example, cupric hydroxychloride [$Cu_2(OH)_3Cl$] and cupric oxide (CuO) have been found in significant amounts under certain conditions[9],

$$Cu_2O_{(s)} + 2H_2O_{(l)} + Cl^{-}{}_{(aq)} \rightarrow Cu_2(OH)_3Cl_{(s)} + H^{+}{}_{(aq)} + 2e^{-} \quad (5)$$

Mansfeld *et al.*[7] reported that the Atacamite is much less protective than cuprous oxide (Cu_2O) in marine environments.

The incidence of failures of aluminium brass heat exchanger tubes and copper nickel iron seawater pipes occurring after only a short service time have been alarming. These have been associated with the presence of sulphide films on the alloy surfaces as a result of a polluted seawater environment. Extensive research has shown that corrosion is particularly aggressive in polluted harbours and estuaries, where the ebb and flow of tides cause a change from polluted anaerobic to relatively fresh aerated seawater[10-15]. When transferring from a polluted to a non-polluted environment, the sulphide and the oxygen can co-exist for a short period. These studies showed that accelerated attack is not caused by the presence of sulphide alone but when sulphide and oxygen contact the metal surface simultaneously.

Syrett[11] has suggested that the sulphide ion (or sulphide oxidation products) can interfere with the normal growth of the protective oxide film that forms on the surface of copper nickel alloys exposed to seawater. The result is the production of a porous, non-protective cuprous sulphide film, which interferes with the normal passivation of the alloy when exposed to unpolluted aerated seawater. According to Sanchez and Schiffrin[16] the main effect of the pollutant is to drastically increase the rate of the oxygen reduction reaction, while the anodic process remains largely unaltered.

A proposed corrosion mechanism for copper nickel alloys in sulphide polluted waters is given below[17]. The cathodic reaction in deaerated seawater,

$$2H^+_{(aq)} + 2e^- \rightarrow H_{2(g)} \quad (6)$$

or in aerated seawater

$$O_{2(g)} + 2H_2O_{(l)} + 4e^- \rightarrow 4OH^-_{(aq)} \quad (1)$$

can be accompanied by,

$$Cu\text{-}Ni_{(s)} + HS^-_{(aq)} \rightarrow Cu\text{-}Ni(HS^-)_{ads} \quad (7)$$

$$Cu(HS^-)_{ads} \rightarrow Cu(HS)_{(s)} + e^- \quad (8)$$

$$Ni(HS^-)_{ads} \rightarrow Ni(HS)^+_{(aq)} + 2e^- \quad (9)$$

$$Cu(HS)_{(s)} \rightarrow Cu^+_{(aq)} + HS^-_{(aq)} \quad (10)$$

$$Ni(SH)^+_{(aq)} \rightarrow Ni^{2+}_{(aq)} + HS^-_{(aq)} \quad (11)$$

The reaction of Cu^+ and Ni^{2+} ions with HS^- can produce copper and nickel sulphides, which are insoluble and precipitate on the alloy surface,

$$2Cu^+_{(aq)} + HS^-_{(aq)} + OH^-_{(aq)} \rightarrow Cu_2S_{(s)} + H_2O_{(l)} \quad (12)$$

$$Ni^{2+}_{(aq)} + HS^-_{(aq)} + OH^-_{(aq)} \rightarrow NiS_{(s)} + H_2O_{(l)}. \quad (13)$$

At the pH of seawater (8-8.2), the sulphide ion is not stable and the hydrosulphide ion is preferred,

$$S^{2-}{}_{(aq)} + H_2O_{(l)} \rightarrow HS^{-}{}_{(aq)} + OH^{-}{}_{(aq)} \quad (14)$$

The OH^- ions produced increase the local pH at the electrode surface and form insoluble hydroxides,

$$Cu^{+}{}_{(aq)} + OH^{-}{}_{(aq)} \rightarrow Cu(OH)_{(s)} \quad (15)$$

$$Ni^{2+}{}_{(aq)} + 2OH^{-}{}_{(aq)} \rightarrow Ni(OH)_{2(s)} \quad (16)$$

The corrosion products tend to be a brittle, non-adherent and non-protective mixture of hydroxides and sulphides of copper and nickel.

In the present work, the oxygen reduction reaction on copper and MARINEL™ in sulphide contaminated and in uncontaminated sodium chloride solutions has been investigated using a rotating disc electrode. An emerging technique, atomic force microscopy has been used to study the initiation of the corrosion process of copper and its alloy, and also to correlate the influence of passive layers on the corrosion reaction.

2. EXPERIMENTAL DETAILS

2.1 Materials

Copper metal, 99.9% pure, was supplied by Goodfellow Metals. MARINEL™, the wrought and precipitate hardened copper nickel, was obtained from Langley Alloys in rod form. Its composition is given in Table 1.

Table 1 Composition of MARINEL™

Element	Mn	Ni	Cu	Al	Nb	Fe	Pb
wt %	4.57	19.1	71.93	1.9	0.72	1.18	0.60

2.2 Electrochemical investigations

Electrochemical measurements were made using an EG&G model 636 rotating disc electrode system coupled to an EG&G Princeton Applied Research 273A Potentiostat/Galvanostat, or an ACM potentiostat and an ACM S300 sweep generator with the current *versus* potential output recorded on a Philips PM 8271 XYt chart recorder. The rotating disc electrode (RDE) consisted of a copper or copper nickel alloy disc of 0.2 cm radius, set in an insulated PTFE shroud of radius 0.6 cm. The experiments were carried out in a three compartment glass cell with the counter electrode (platinum gauze) separated from the working electrode by a Nafion™ cation exchange membrane. All electrode potentials were measured with respect to a saturated calomel electrode (Radiometer K410). The pH of the solutions was monitored using a Jenway 3420 electrochemistry analyser.

The RDEs were polished with a 0.3 μm Al_2O_3/water slurry. Prior to each polarisation measurement the sodium chloride solutions were aerated for 5 min to produce an air saturated electrolyte. The rest potential of the working electrode was measured with respect to the SCE, and then the potential was linearly scanned approximately 1000 mV in the negative direction. For the diffusion coefficient measurements, polarisation curves were obtained for rotation rates ranging from 250 to 2000 rev min^{-1} at a potential sweep rate of 2 mV s^{-1} to maintain steady state conditions.

2.3 Atomic force microscopy

Images were obtained in the contact mode with a Discoverer TMX 2010 Atomic Force Microscope (Topometrix Corporation, Santa Clara, California, USA); the force constant was 0.012 nN.

The copper nickel alloy was cut into 10 mm diameter, 2 mm thick discs. The samples were polished with an 0.3 μm alumina/water slurry, washed with double distilled water and finally rinsed with absolute alcohol. The air-dried samples were placed in the solutions of interest for varying time intervals after which they were removed, dried in air and observed directly.

3. RESULTS AND DISCUSSION

3.1 Electrochemical investigations

Figures 1a-b show polarisation curves for oxygen reduction, equation (1), on copper and copper nickel in aerated, 3% sodium chloride solution (pH = 6.2 and 293 K). In both cases, well-defined plateau regions, due to convective diffusion limited oxygen reduction, are observed. These extend from -650 to -1150 mV (vs.SCE), at which point the onset of the hydrogen evolution reaction can be observed.

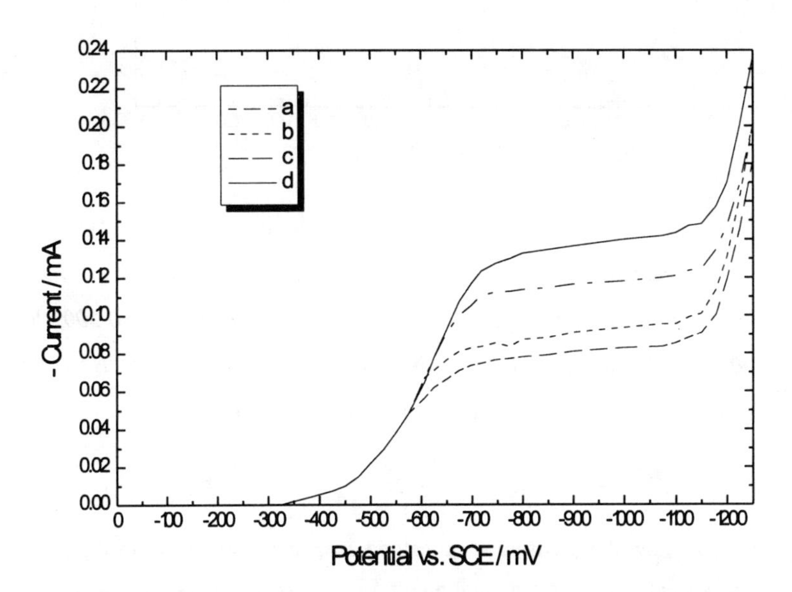

Figure 1a *Current-potential curves for the reduction of oxygen in aerated 3% sodium chloride solution at a copper RDE. Rotation rates were (a) 500, (b) 750, (c) 1250 and (d) 1750 rpm. WE: copper disc, RE: SCE and CE: Platinum. Potential sweep rate 2 mV s^{-1}, pH = 6.2 and T = 293 K.*

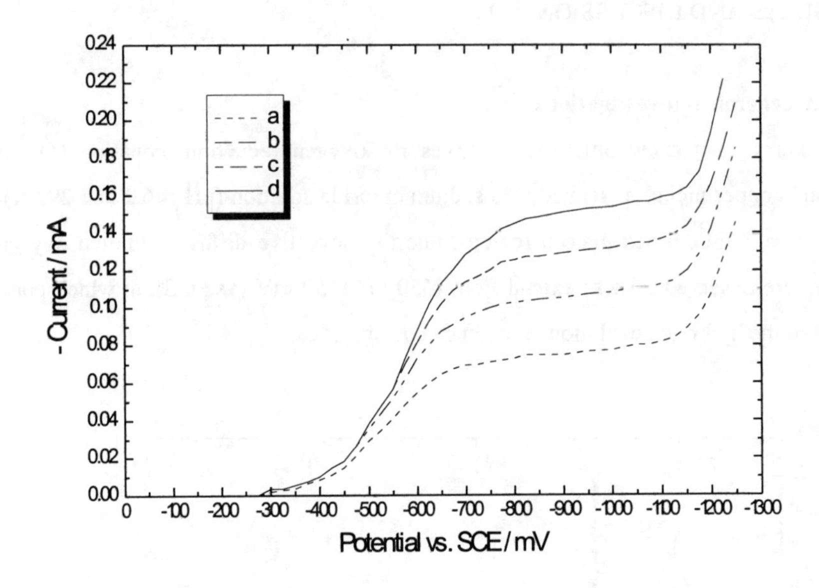

Figure 1b *Current-potential curves for the reduction of oxygen in aerated 3% sodium chloride solution at a copper nickel RDE. Rotation rates were (a) 500, (b) 1000, (c) 1500 and (d) 2000 rpm. WE: copper nickel disc, RE: SCE and CE: Platinum. Potential sweep rate 2 mV* s^{-1}, *pH = 6.2 and T = 293 K.*

Figures 2a-b show the corresponding cathodic polarisation curves in the presence of 5 ppm sodium sulphide (pH 7.2 and 293 K). A second limiting current region extending from -480 to -850 mV (SCE) is observed for both copper and copper nickel. However this second plateau is less well defined on the alloy surface.

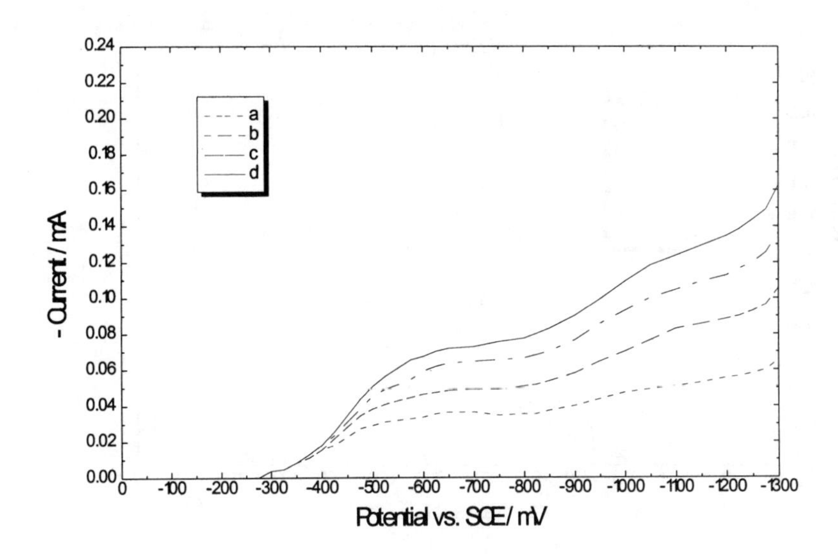

Figure 2a *Current-potential curves for the reduction of oxygen in aerated 3% sodium chloride containing 5 ppm sodium sulphide, solution at a copper RDE. Rotation rates were (a) 250, (b) 750, (c) 1250 and (d) 1750 rpm. WE: copper disc, RE: SCE and CE: Platinum. Potential sweep rate 2 mV s^{-1}, pH = 7.2 and T = 293 K.*

Levich plots of limiting current *versus* the square root of the rotation rate (Figure 3) were linear, showing the oxygen reduction reaction to be mass transport controlled. From this data, the diffusion coefficient for oxygen reduction at the RDE under steady state conditions was calculated *via* the Levich equation (17).

$$I_L = 0.62\, n\, F\, A\, D^{2/3}\, \nu^{-1/6}\, \omega^{1/2}\, c_B \qquad (17)$$

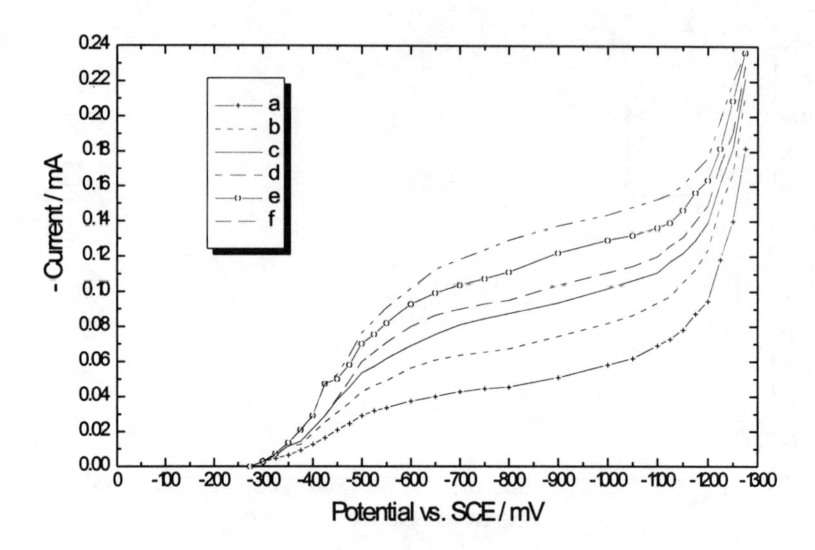

Figure 2b *Current-potential curves for the reduction of oxygen in aerated 3% sodium chloride containing 5 ppm sodium sulphide, solution at a copper nickel RDE. Rotation rates were (a) 500, (b) 1000, (c) 1250, (d) 1500, (e) 1750 and (f) 2000 rpm. WE: copper nickel disc, RE: SCE and CE: Platinum. Potential sweep rate 2 mV s^{-1}. The pH was 7.2 and T = 293 ± 2 K.*

Assuming the cathodic reaction is the reduction of oxygen, (n = 4) and the bulk concentration of oxygen (c_B) is 2.0 x 10^{-4} mol dm^{-3} [18], the diffusion coefficients for oxygen reduction on copper nickel and copper rotating disc electrodes were calculated to be 1.50 x 10^{-5} cm^2 s^{-1} and 1.44 x 10^{-5} cm^2 s^{-1} , respectively. These values compare favourably with those of 1.4 x 10^{-5} cm^2 s^{-1} obtained by Wood *et al.* [19] for 70/30 copper nickel, (B.S.2875), and 1.7 x 10^{-5} cm^2 s^{-1} for aluminium brass in natural seawater[16].

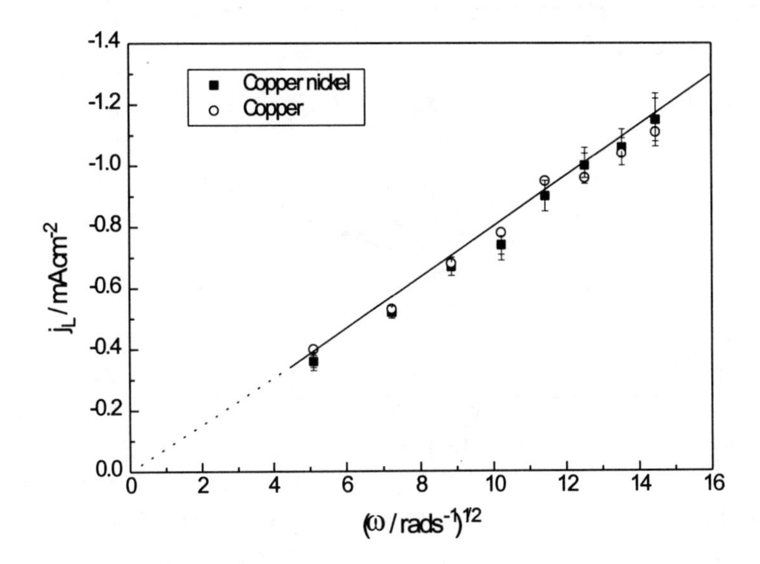

Figure 3 *Levich plots for copper and copper nickel RDEs in aerated 3% sodium chloride at pH = 6.2 and 293 K.*

Levich plots for the copper nickel and copper in aerated 3% sodium chloride solutions containing 5 ppm sodium sulphide are shown in Figures 4 and 5, where slope (b) represents the limiting current plateau between -480 and -850 mV. The values of the diffusion coefficients for oxygen reduction, slope (a), were slightly reduced to 1.35 x 10^{-5} and 1.24 x 10^{-5} cm^2 s^{-1} for the copper nickel and copper respectively. This may result from the reaction of sulphide with the dissolved oxygen as proposed by Beccaria *et al.* [20]. The chemistry of aqueous sulphides and oxidation products is extremely complex but available data indicates that dissolved sulphide is oxidised fairly rapidly in both sodium chloride and seawater solutions. Ostlund and Alexander[21] showed that for air saturated seawater with a sulphide concentration of 3.8 ppm, the half-life of the sulphide was of the order of 20 min. Avrahami *et al.*[22], in a detailed mechanistic study of this reaction, showed that elemental sulphur, sulphite (SO_3^{2-}), sulphate (SO_4^{2-}) and thiosulphate ($S_2O_3^{2-}$) are formed and that the oxidation of sulphur to the oxyanions is slow. Also sulphides may be oxidised by dissolved oxygen to a variety of polysulphides.

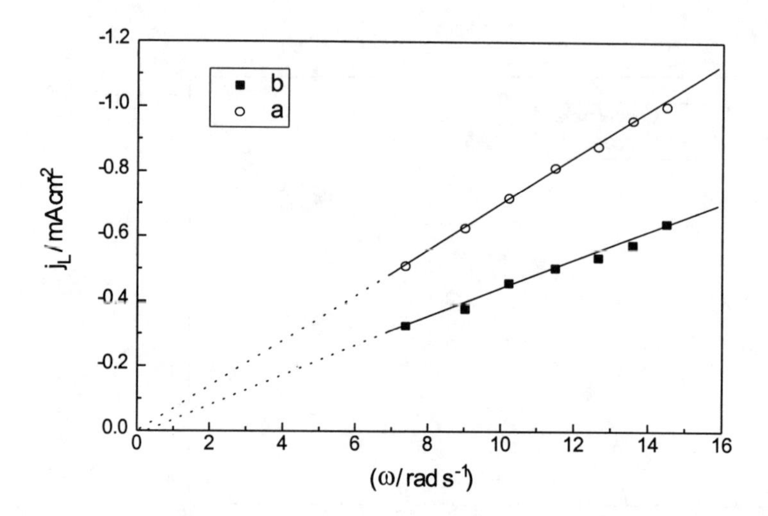

Figure 4 *Levich plots for a copper RDE in aerated 3% NaCl + 5 ppm Na_2S at pH 7.2 and 293 K. (a) Represents oxygen reduction plateau at -650 to -1150 mV and (b) the additional plateau at -480 extending to -850 mV.*

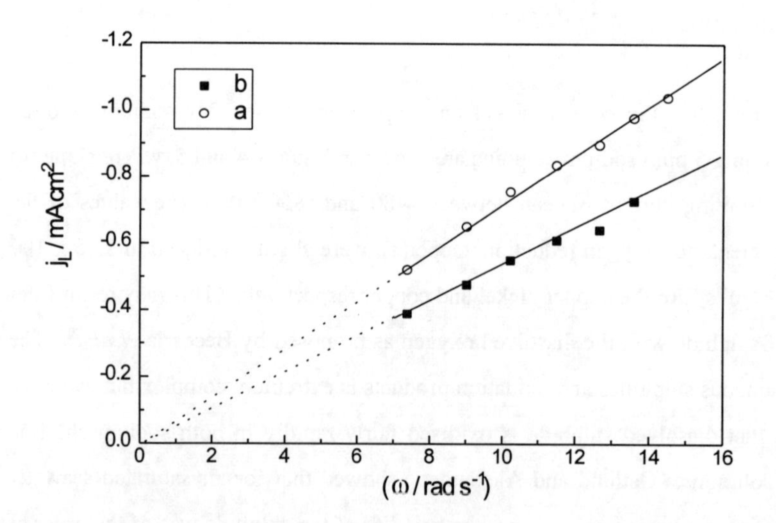

Figure 5 *Levich plots for a copper nickel RDE in aerated 3% NaCl containing 5 ppm Na_2S, pH 7.2, 293 K. (a) represents oxygen reduction plateau at -650 to -1150 mV and (b) the additional plateau at -480 extending to -850 mV.*

The magnitude of the slope for the second plateau region is smaller than that for the reaction (1). It is possible that it is due to the production of hydrogen peroxide[23],

$$O_{2(g)} + 2H_2O_{(l)} + 2e^- \rightarrow H_2O_{2(aq)} + 2OH^-_{(aq)} \quad (18)$$

$$H_2O_{2(aq)} + 2e^- \rightarrow 2OH^-_{(aq)}. \quad (19)$$

The additional peaks observed in the presence of sulphide ions can be explained by a mechanism involving oxygen reduction on two types of surface sites with different reactivities[23].

It has been reported that the more catalytic surface is comprised of Cu(0) and Cu(I) sites, where the Cu(I) species is stabilised as $Cu(OH)_{ads}$ and/or submonolayer Cu_2O. The less catalytic site consists of Cu(0) only. Oxygen reduction is believed to proceed by a series pathway involving an adsorbed peroxide intermediate on both sites. The majority of the peroxide is reduced prior to desorption at Cu(0) sites. However, some is released before being reduced at Cu(0)/Cu(I) sites, hence, the presence of the additional peak at -400 mV *vs.* SCE.

This was confirmed by the results of King *et al.*[23] who studied the oxygen reduction on copper in neutral sodium chloride (1 mol dm^{-3}, T= 296±3 K) using a rotating ring disc electrode. They found that the extent of the surface coverage by the catalytic species was favoured by a higher interfacial pH, which is in turn favoured by a higher electrolyte pH. In the presence of sulphide, the electrolyte pH increases from a slightly acidic pH 6.2 to 7.2.

They derived a scheme for oxygen reduction on the two types of copper surface sites. The less catalytic surface site was designated S_A and here the oxygen reduction proceeded with no change in the oxidation state of the Cu(0) species. They proposed that the reaction proceeds via the adsorption of a peroxide (HO_2^-) intermediate on the surface, which is reduced prior to desorption as in shown in equation (20).

$$S_A + O_2 \rightarrow S_A\text{-}O_{2(ads)} \xrightarrow{+e} S_A\text{-}O_2^-{}_{(ads)} \xrightarrow{+e+H^+} S_A\text{-}HO_2^-{}_{(ads)} \rightarrow S_A + 4OH^- \quad (20)$$

$$S_B + O_2 \rightarrow S_B\text{-}O_{2(ads)} \xrightarrow{+e} S^+{}_B\text{-}O_2^-{}_{(ads)} \rightarrow S_B\text{-}O_2^-{}_{(ads)} \xrightarrow{+H^+} S^+{}_B\text{-}HO_2^-{}_{(ads)} \xrightarrow{+e} S_B\text{-}HO_2^-{}_{(ads)} \begin{cases} \nearrow HO_2^-{}_{(bulk)} \\ \swarrow S_B + 4OH^- \end{cases} \quad (21)$$

where S_A = Cu(0) (less catalytic) and S_B = Cu(0)/Cu(I) (catalytic). The more catalytic site S_B exists as either Cu(0) or Cu(I) species. The Cu(I) surface site is stabilised by the adsorption of OH^- (or Cl^-) or as a submonolayer Cu_2O. Oxygen reduction proceeds via electron relay between Cu(0)/Cu(I) sites and adsorbed oxygen species. Unlike S_A sites, the rate of reduction of the peroxide does not greatly exceed the rate of desorption and some peroxide is released into solution. The percentage surface coverage by the two types of site is determined by the interfacial and electrolyte pH and potential. Higher interfacial values (and higher bulk pH) favour the formation of the more catalytic site. The results of the present study are consistent with this theory as no additional peroxide peak was produced in the absence of sulphide where a more acidic pH of 6.2 exists.

The theory is further supported by calculating the number of electrons involved in the two reduction processes on the copper and copper nickel via the Levich equation (utilising the diffusion coefficients determined in the absence of sulphide). For copper, the number of electrons involved changes from 2.3 to 3.8, effectively 2 to 4. For the copper nickel, the change is from 2.9 to 3.8 effectively 3 to 4. Hence, the intermediate formation of peroxide, reaction (18), is the two electron process producing the additional plateau. If the peroxide intermediate is strongly adsorbed (and reaction (19) is fast) prior to desorption, the reduction of oxygen is a 4 electron process. However, if it is not strongly adsorbed, the stoichiometry will depend on the diffusional conditions imposed[19]. Thus, the total number of electrons involved alters from 4 to 2 as the flow velocity increases and less peroxide is reduced. Hence, the second plateau becomes more prominent at higher rotation rates of a disc electrode.

3.2 Atomic force microscopy investigations

It has been stated that the protective corrosion products form within minutes or even seconds on freshly polished copper nickel alloys[11]. To investigate this, atomic force microscopy was used to study the morphology and rate of film formation on copper nickel after exposure to aerated 3% sodium chloride and sodium sulphide contaminated solutions.

Figure 6 shows an *ex situ* AFM image of the unexposed copper nickel surface. Scratch marks from the polishing process and particles of the alumina itself can be clearly seen on the sample surface. Subsequent investigations showed that the alumina did not interfere or contribute to the surface film formation as there was no correlation between the location of the corrosion product nucleation sites and the alumina present. However, to avoid any possible problems, later samples were sonicated prior to exposure to the relevant electrolyte.

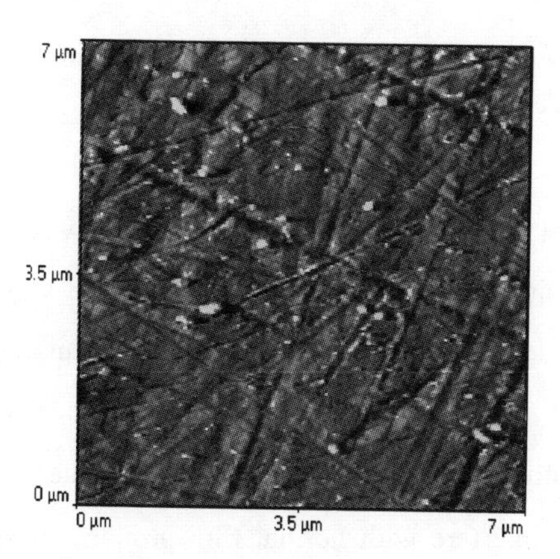

Figure 6 *Atomic force microscopy image of polished copper nickel alloy showing traces of alumina on the metal surface.*

Film formation on the copper nickel alloy in aerated 3% sodium chloride solution, as seen in Figures 7a and b, was rapid. After a 5 minute exposure to the solution, a particulate film is evident covering the majority of the sample surface with only one polished area remaining visible. However, this region was obscured after 60 minutes when the surface of the copper nickel alloy was seen to be completely covered with a uniform layer.

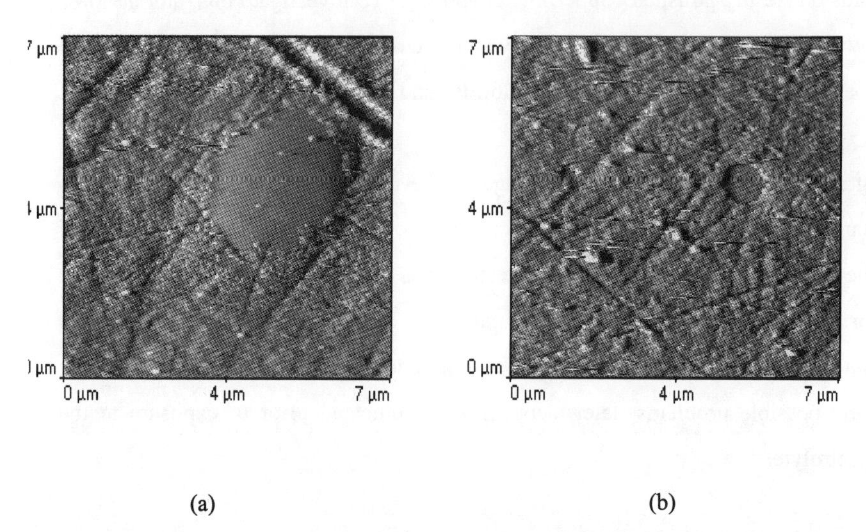

(a) (b)

Figure 7 *Copper nickel exposed to aerated 3% sodium chloride solution for (a) 5 minutes and (b) 60 minutes.*

Line profiles of the unfilmed area in Figure 7a showed it to be a convex protrusion from the surface. This suggests that it is most likely to be a precipitate known to form in aluminium age hardened copper nickels. Such micron scale features ranging from 0.1 to 3 µm diameter have been found in MARINEL™ which analysis has confirmed as Ni_3Al, Ni_3Nb, $Ni_{16}NbSi_9$ and (more commonly) orthorhombic Ni-Nb-Si-Fe intermetallic phases[24]. It is feasible that these particles interfere with normal film growth. Scanning electron microscopy coupled with energy dispersive analysis of the sample proved inconclusive, as this is a *near* surface technique and the electron beam can penetrate the surface to depths of approximately 1 µm[25].

In the presence of 5 ppm sulphide, a uniform film formed on the copper nickel sample after 5 minutes exposure, Figures 8a-b. This suggests more rapid film formation in the presence of sulphide ions. In both solutions, the films continued to grow with time as indicated by the disappearance of the scratch marks originally present from the polishing process.

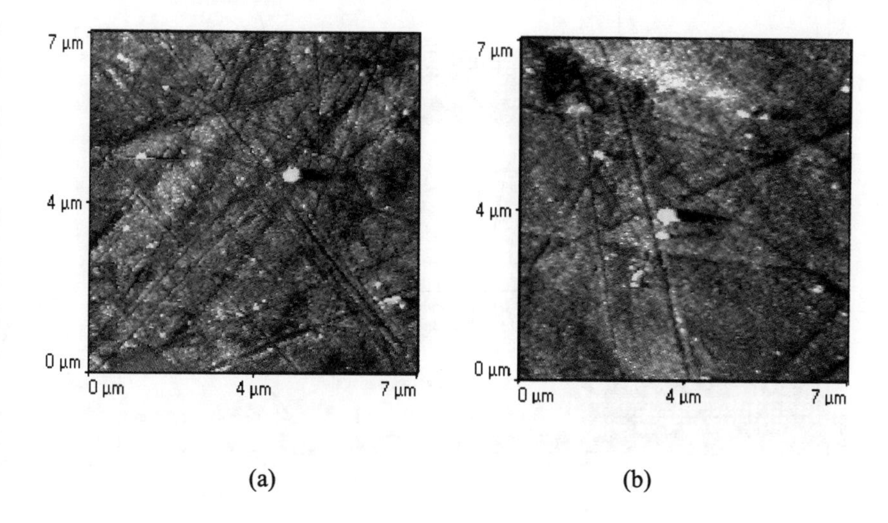

Figure 8 *Copper nickel exposed to aerated 3% sodium chloride solution containing 5 ppm sodium sulphide for (a) 5 minutes and (b) 60 minutes.*

These results agree with the proposal that within the first few seconds of exposure to sulphide polluted seawater, when the corrosion potential still exceeds *ca.* -260 mV (vs.SCE), a thin film or sub-layer forms on the metal surface (deduced from scanning electron microscopy and X-ray diffraction data) [11]. This film, termed the oxide-type-layer (Figure 9), is thought to be predominately cuprous oxide doped with the alloying elements (Ni, Fe and possibly Mn). Due to the composition and microstructural variations at grain boundaries, the oxide does not grow as fast at these locations, therefore, these boundaries appear as cracks in the corrosion product.

The typical alloy grain size in the copper nickel, shown in Figure 10, ranges from approximately 40 μm^2 to 100 μm^2, which is much larger than the scan range (7 µm × 7 µm) of the AFM micrographs shown. Therefore, it is reasonable to assume that the AFM images show the growth of the particulate layers on top of such a sub layer, as shown in Figure 9.

(a)

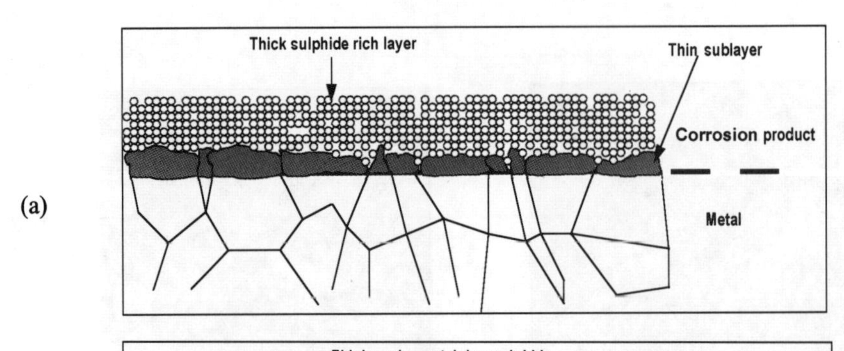

(b)

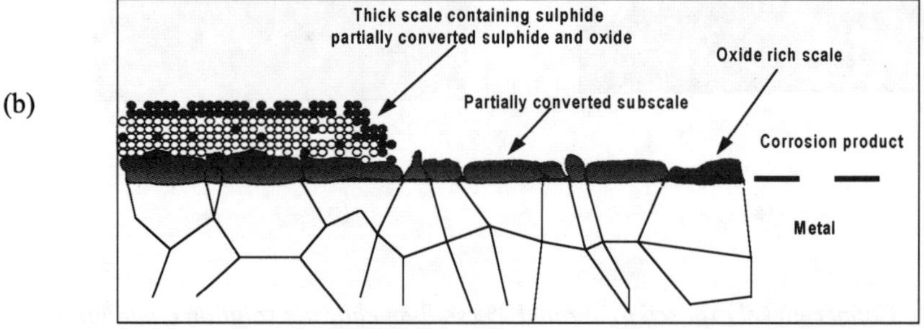

Figure 9 *Schematic diagrams showing the corrosion products resulting from exposure to (a) sulphide polluted seawater and (b) sulphide polluted deaerated seawater then aerated clean seawater*[11].

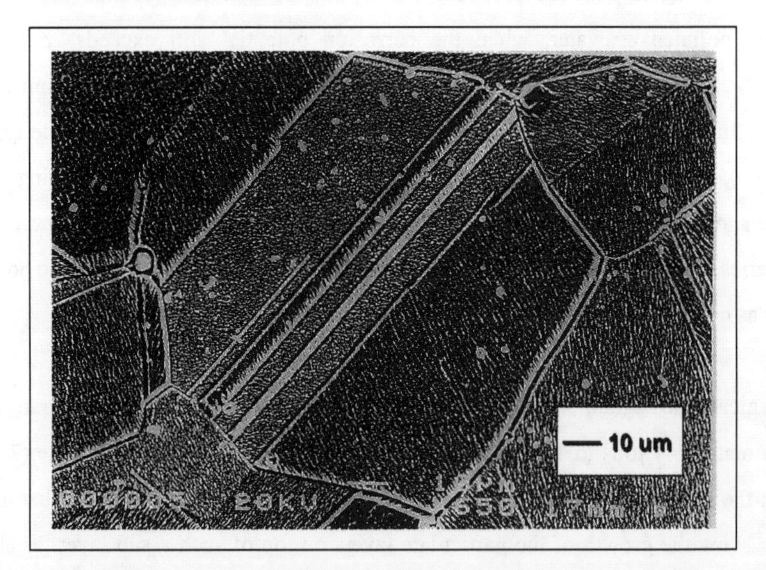

Figure 10 *Scanning electron micrograph of ferric chloride etched MARINEL™ showing the alloy grain boundaries.*

The images shown in Figure 11 show that the particulate sizes of the layers formed on the samples, varied for the contaminated and uncontaminated solutions. This can be clearly seen by the topographic analyses in Figures 12 and 13.

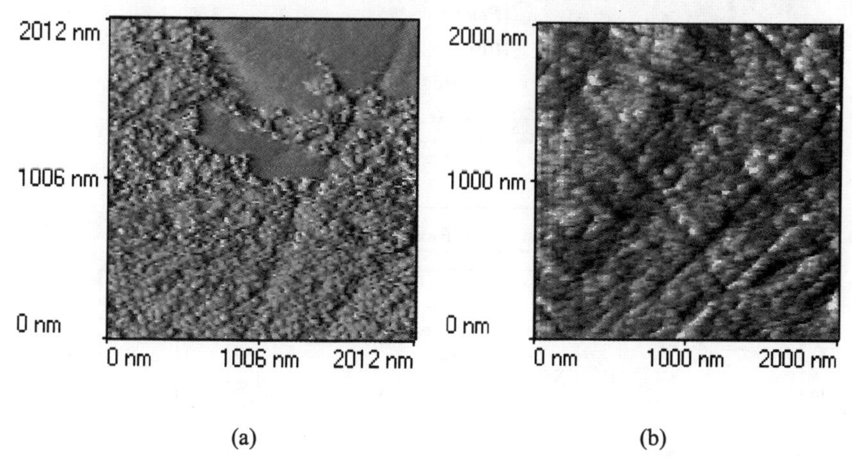

(a) (b)

Figure 11 *Copper nickel exposed to (a) uncontaminated and (b) contaminated (with 5 ppm sodium sulphide) 3% sodium chloride solution (aerated) for 5 minutes.*

The particulates formed in both solutions appear to have a "granular" type structure. In the presence of sulphide pollution, the average film particle size was 300 nm in diameter, compared to 68 nm for the sulphide-free sea water. However, similar particle heights of 6 nm were observed for both types of particulates.

The increase in the particle size of the surface films could account for the well-documented increase in porosity[10-15] of the film towards the corrosive environment, hence reducing its corrosion resistance.

This result is somewhat unexpected as according to the theory of oxidation, the film formed in the presence of sulphide contamination should be more protective. However, this does not account for the more defective structure of metal sulphides, film spalling and the enhanced cathodic reduction reaction occurring on the sulphide layer, resulting in the pitting corrosion[26]. The oxidative description of protective film formation

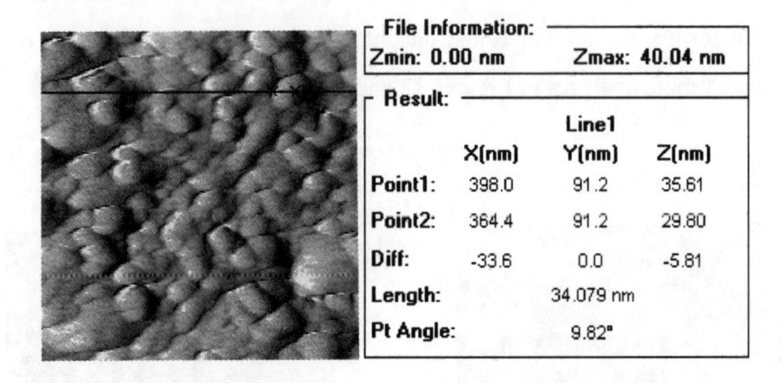

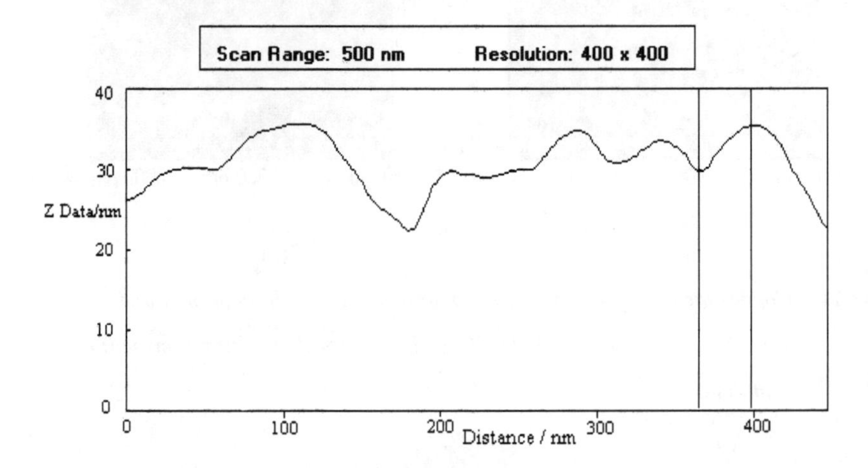

Figure 12 *Topographic analysis of copper nickel exposed to aerated 3% sodium chloride for 5 minutes.*

suggests an eventual parabolic corrosion rate governed by diffusion of a cation *via* vacancies, which rather implies lattice diffusion. This is obviously an oversimplification. The films formed consist of grains and preferential diffusion down grain boundaries may be a major feature of the growth process. Such mechanisms place great emphasis upon the microstructure of the oxide particularly the grain size. It could be assumed, therefore, that the finer grains formed in the unpolluted solution would result in a relatively high density of grain boundary paths for oxide growth when compared to the film formed in sulphide contaminated solution.

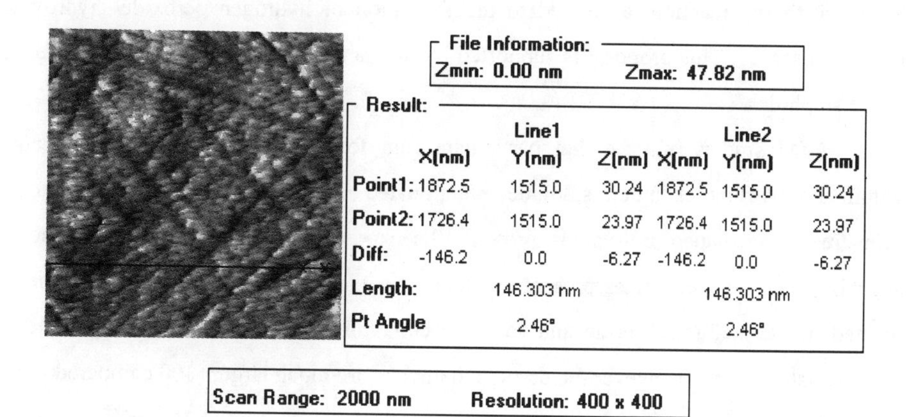

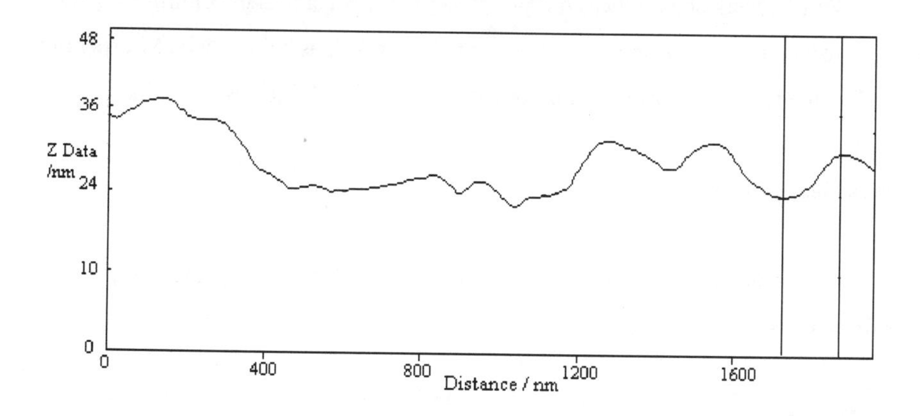

Figure 13 *Topographic analysis of copper nickel surface exposed to aerated 3% sodium chloride containing 5 ppm sodium sulphide for 5 minutes.*

4. CONCLUSIONS

Electrochemical studies of copper and MARINEL™ in seawater show that the limiting current for oxygen reduction at both electrodes is mass transfer controlled, due to the rate of convective diffusion of dissolved oxygen. The reaction itself is similar on both copper nickel and copper substrates with comparative limiting currents at potentials of -650 to -1150 mV (vs.SCE) from which diffusion coefficients of 1.50×10^{-5} and 1.44×10^{-5} $cm^2\ s^{-1}$ respectively may be calculated. In the presence of sulphide ions, a second mass transport limited cathodic reaction, attributed to the formation of hydrogen peroxide/ hydroxide ions, is observed. This proposal is supported by the increased solution pH in the presence of the sulphides.

AFM studies indicate that particulate film formation on copper nickel occurs within about 5 minutes in both sulphide contaminated and pure sodium chloride solutions. Film-free dome shaped protrusions, present after exposure to uncontaminated conditions, are thought to be precipitates present in the copper nickel. The heights of the particles formed in the sodium chloride and sulphide contaminated solutions are comparable at approximately 6 nm. However the particle diameters are much larger (300 compared to 68 nm) in the presence of sulphide.

Further work should focus on spectroscopic investigations to confirm the nature of the products formed at the more negative potentials as well as on *in situ* AFM experiments to follow the very early stages of film formation in a variety of corrosive media.

Acknowledgements

The authors would like to thank the MTD for the EPSRC studentship and Langley Alloys for financial support.

References

1. C. A. Clark, S.Driscoll and P. Guha, *British Corrosion Journal*, 1992, **27**, 157.
2. G. Bianchi, G. Fiori, P. Longhi and F. Mazza, *Corrosion*, 1978, **34**, 396.
3. A.H. Tuthill, *Materials Performance*, 1987, **12**, 12.
4. R.F. North and M.J. Pryor, *Corrosion Science*, 1970, **10**, 297.
5. J..M.Popplewell, R. J. Hart and J. A. Ford, *Corrosion Science*, 1973, **13**, 295.
6. C. Kato, J. E. Castle, B. G. Ateya and H. W. Pickering, *J.Electrochem.Soc.*, 1980, **127**, 1897.
7. F. Mansfeld, G.Liu, H. Xiao, C.H Tsai and B.J Little, *Corrosion*, 1994, **36**, 2064.
8. S. R.de Sanchez, *J. Electroanal. Chem.*, 1991, **307**, 73.
9. J. Kruger, *J.Electrochem.Soc.*, 1961, **108**, 503.

10. J. F. Bates and J. M Popplewell, *Corrosion NACE*, 1975, **31** No 8, 269.
11. B. C. Syrett, *Corrosion Science*, 1981, **21**, 187.
12. B. C. Syrett, D. D. Macdonald and S.S.Wing, *Corrosion, NACE*, 1979, **35**, 409.
13. L.E Eiselstein, B.C Syrett, S.S. Wing and D.Caligiuri, *Corrosion Science*, 1983, **23**, 223.
14. B.C Syrett, *Corrosion NACE*, 1977, **23**, 257.
15. J.P. Gudas and H.P Hack, *Corrosion NACE*, 1979, **35**, 67.
16. S.R. de Sanchez and D.J Schiffrin, *Corrosion Science*, 1982, **22**, 597.
17. M.R.Reda and N.Alhajji, *Br. Corr. J.*, 1995, **30** No.1, 58.
18. M. Stratmann and J. Muller, *Corrosion Science*, 1994, **36**, 327.
19. R.J.K. Wood, S.P. Hutton and D.J.Schiffrin, *Corrosion Science*, 1990, **30**, 1199.
20. A.M.Beccaria, G. Poggi, P. Traverso and M. Ghiazza, *Corrosion Science*, 1991, **32**, 1267.
21. H. G. Ostlund and J. Alexander, *J. Geophys. Res.*, 1963, **68**, 3995.
22. M. Avrahami and R.M. Golding, *J. Chem. Soc.*, (1968), 647.
23. F.King, M.J.Quinn and C.D Litke, *J. Electroanal. Chem.*, 1995, **385**, 45.
24. C.D.S Tuck, K.C. Bendall, G.J.W. Radford, S.A Campbell, F.C. Walsh and R.J. Grylls, NACE International Annual Conferrence Proceedings, 1996.
25. R. Baboian, Corrosion Tests and Standards, Application and Interpretation, ASTM Manual Series, 1995,
26. G.J.W. Radford, The Marine Corrosion and Electrochemical Characteristics of MARINEL Copper Nickel Alloy, PhD Thesis, (1998), 235.

4. THE USE OF *IN-SITU* INFRARED SPECTROSCOPY AS A PROBE FOR INVESTIGATING CORROSION REACTIONS

John F. Halsall, Maher Kalaji (for correspondence) and L. Keri Warden-Owen

Department of Chemistry, University of Wales at Bangor, Bangor LL57 2UW, UK

1. INTRODUCTION

The use of *in-situ* infrared spectroscopy to study reactions at the surfaces of electrodes was first reported by Bewick and co-workers in 19791[1]. Since then, the original technique, EMIRS (electrochemical modulated infrared spectroscopy), and other variants, based on grating or FTIR spectrometers have been successfully used to elucidate a large number of reaction mechanisms occurring at metal or semiconductor electrodes.

In this work, we report the results of experiments using subtractively normalised interfacial FTIR spectroscopy (SNIFTIRS) on iron and steel electrodes in aqueous media in the presence of bisulphate and sulphate anions. The experiments are the first in a programme which aims to provide an understanding of the processes occurring at the metal/biofilm/electrolyte interphases that lead to the corrosion of metals in marine environments.

1.1 The Passivation Behaviour of Stainless Steel (SS304) in Sulphuric Acid

The composition and thickness of the passive film formed on SS304 has been shown to be related to the potential at which it was passivated (Figure 1). At low potentials in the passive region the composition mainly consists of chromium oxide; at more positive potentials the surface becomes a more complex mixture of oxides of the alloy constituents. Close to the transpassive region the film is thought to be mainly iron oxide, its formation being associated with the dissolution of the other alloying components.

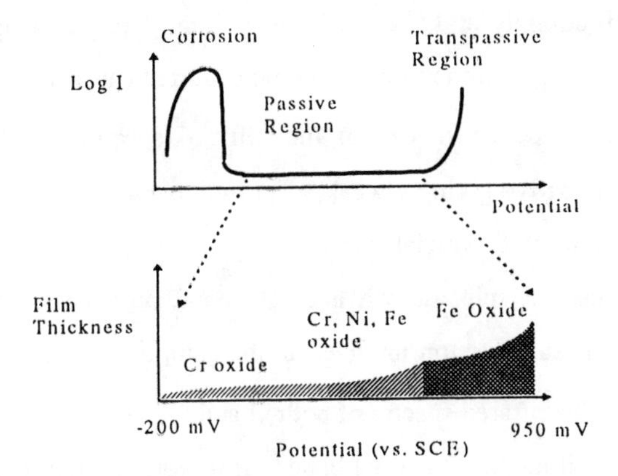

Figure 1 *Composition of SS304 Stainless steel in 1 mol dm^{-3} Sulphuric Acid*

1.2 Infrared Characteristics of Sulphates

The infrared characteristics of sulphates are relatively well known[2]. The free sulphate ion (SO_4^{2-}) is highly symmetrical being a member of the T_d point group. Some of the possible vibrational modes associated with this anion are shown in Figure 2. The sulphate ion has four fundamental modes of vibration of which only ν3 and ν4 are infrared active.

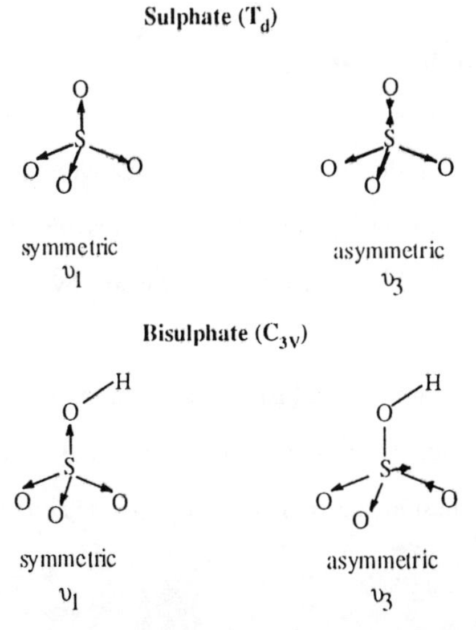

Figure 2 *Infrared Vibrational Modes of Sulphate and Bisulphate*

The ν3 and ν4 bands adsorb strongly at 1104 cm^{-1} and 613 cm^{-1} respectively. When sulphate ions are complexed, the T_d symmetry can become distorted causing a degenerate vibrational split and activating Raman modes in the infrared spectrum. When this happens, the ν1 mode becomes active giving a weak adsorption band at around 970 cm^{-1} due to the interaction of the anions with a metal surface.

The bonding of a metal to sulphate via a single bond produces a unidentate complex. The symmetry of the sulphate ion involved in the complex is reduced to C_{3V}. This allows the appearance in the infrared spectra of both νl and ν2 bands (at 450 cm^{-1}) as well as causing the two way splitting of ν3 and ν4 bands. It is worth noting that both the ν2 and ν4 modes occur beyond the detection limit of our present detector system.

If the sulphate is bound by two bonds to the metal the symmetry is reduced to C_{2V}. This bidentate complex results in the activation in the infrared region of the νl and ν2 Raman bands and the three way splitting of the ν3 and ν4 bands. Furthermore, two types of bidentate exist. The first, bridged bidentate has two metal atoms linked by the sulphate ion. The second, chelated, one has a single metal ion doubly bonded to a sulphate ion. The infrared spectra of the two complexes are similar but they can be distinguished as the chelated form has higher band frequencies than for the bridged.

1.3 Infrared Characteristics of Bisulphate

The C_{3V} symmetry is approximately the same as that of unidentate bonded sulphate (see above) with a hydrogen atom rather than a metal (see above). Both the νl and ν3 absorb strongly at 1050 cm^{-1} (symmetric) and 1195 cm^{-1} (asymmetric), respectively. Furthermore, the presence of the S-0-H fragment[4] is indicated by an absorption band, which appears at 900-870 cm^{-1}.

Care must be taken when dealing with bisulphate solutions as the dissociation of bisulphate to sulphate is pH dependant with a pK value of 1.02×10^{-2}. Hence, bisulphate dominates in acidic solutions while sulphate is the majority species at neutral pH.

2. EXPERIMENTAL DETAILS

The working electrodes (iron and steel discs, 9.5 in diameter) used in the electrochemical and spectroelectrochemical measurements were polished using successively finer grades of alumina (1, 0.3 and 0.05 μm). The electrodes were ultrasonically cleaned and rinsed in high purity water before being dried under nitrogen. For the SNIFTIRS experiments, the iron (99.99% Goodfellow) and steel (type 304, Goodfellow) discs were imbedded into an adapted PTFE block using epoxy resin (RS Components). Heat shrink tubing (RS Components) was used to connect the PTFE holding the electrode to a glass syringe plunger, effectively making a flexible joint. The solutions used were of high purity analytical grade chemicals and were degassed with nitrogen for 20 minutes before the start of each experiment. A platinum sheet was used as a secondary electrode. All the potentials are quoted against the saturated calomel electrode (SCE).

SNIFTIRS measurements were performed using an FTIR spectrometer (Bruker IFS 113v) modified for electrochemical measurements. A silicon disc (2.5 mm thick, 25.4 mm diameter) was used as the infrared transparent window separating the electrochemical cell from the evacuated spectrometer. The data, presented as difference spectra, were obtained by collecting spectra, I_1 and I_2, at two different potentials E_1 and E_2 or as a function of time, T_1 and T_2. These reflection spectra are presented as difference spectra by subtracting two spectra (I_2-I_1) and dividing by the background absorption, that is I_1 minus the spectrum taken when the electrode is pulled away from the silicon window (I_w). This ratio, called the change in reflectance, $\Delta R/R = (I_2-I_1)/(I_1-I_w)$, is presented here as the percentage change in transmittance.

Throughout these measurements, 200 interferograms were collected at each potential. Both positive and negative bands can appear in difference spectra, the former representing a decrease in the concentration of the species responsible for that absorption at E_2 compared to E_1, whereas the latter indicates an increase in the concentration of the corresponding species at E_2. The spectra were obtained using a 'staircase' method with each spectrum normalised relative to the single reference spectrum I_1, taken at the start of each experiment. This method was used in preference to the "block" mode as the formation of corrosion products on the surface prevents the repeated measurement of I_2, and I_1.

The data collection and cell potential were controlled using in-house software running through OPUS 2 control programme (Bruker) under an 0S2 operating system on a Personal Computer (Dell DX-50). A digital to analogue converter (Amplicon, PC30AT) was used to control the potentiostat from the PC. Spectra were taken in 50 mV steps, with a 1 minute delay at each potential to allow the surface of the electrode to stabilise.

3. RESULTS AND DISCUSION

Due to the complexity of the SS304 surface, a pure iron electrode was initially used as a model system to provide infrared data on the dissolution of iron in sulphuric acid. Previously, SNIFTIRS has been successfully used to investigate the behaviour of pure iron in alkaline solution[5]. More recent studies in borate buffer (pH 8.4) solutions using surface enhanced Raman spectroscopy (SERS) have provided information about the type of oxides involved[6].

Figure 3(i) shows the SNIFTIRS spectra, normalised relative to -800 mV, obtained during an anodic polarisation from -800 mV to -300 mV. The largest negative band is that centred around 1100 cm^{-1} and corresponds to the ν3 vibration of the sulphate anion. The latter is accompanied by a smaller negative band at 980 cm^{-1}. This is more likely due to the νl stretch, which is normally inactive in the highly symmetrical free sulphate ion but is known to become active where the symmetry of the ion is partially reduced. The positive band at 1200 cm^{-1}, associated with asymmetric bisulphate adsorption, seems to show a decrease in its' concentration in the thin layer. An explanation could be that bisulphate ions react to produce sulphate ions in the double layer, possibly due to an increase in pH. The reduction in symmetry of the sulphate ion could he caused by the spatial impingement caused by its interaction with the Fe^{2+} ion formed as a result of the anodic oxidation of the iron surface. The assignment of these bands to solution rather than surface species has been made because the change in absorbance is more than one order of magnitude higher than that of adsorbed species previously found by this technique[7]. Furthermore, the dissolution of Fe^{2+} ions is expected to occur at these potentials in such a solution .

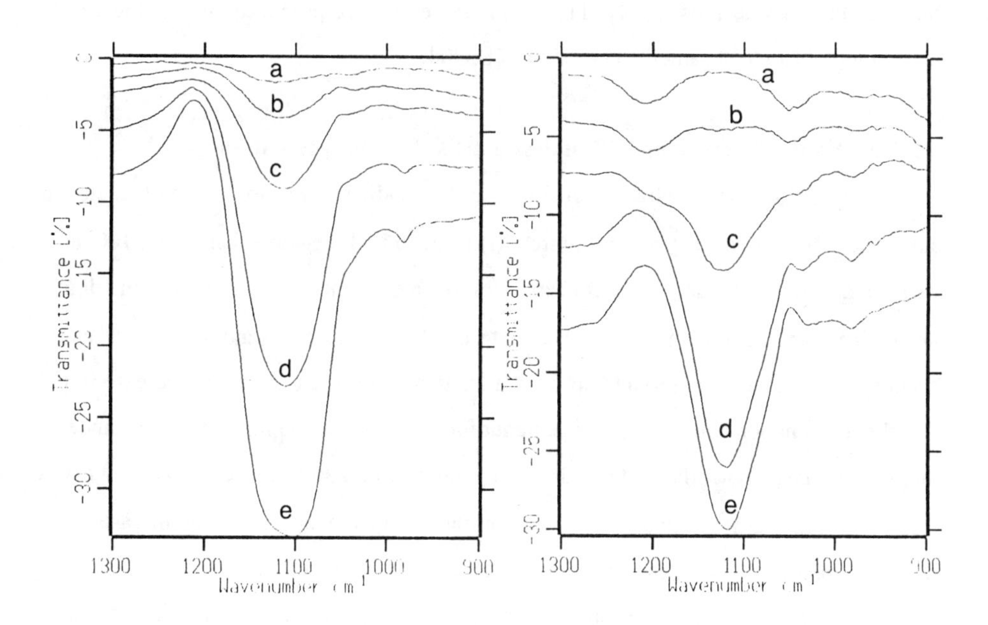

i) Anodic polarisation of iron

Spectra a, b, c, d and c correspond to

E_2 *= -700 mV to-300 mV*

in 100 mV steps

E_1 *= -800 mV*

ii) Cathodic polarisation of SS304.

Spectra a, b, c, d and e correspond to

E_2 *= -100 mV to -500 mV*

in 100 mV steps

E_1 *= 0 mV*

Figure 3 *SNIFTIRS Spectra, obtained during the Polarisation of Iron and Stainless steel (Type 304) in 1 mol dm* $^{-3}$ *Sulphuric Acid*

The infrared behaviour of SS304, in the potential region of high corrosion (<-200 mV in 1 mol dm^{-3} sulphuric acid), closely resembles that of iron. Figure 3(ii) shows the cathodic polarisation of SS304 between -100 mV and -500 mV normalised relative to 0 mV. The negative band at 1100 cm^{-1} (ν3, sulphate) begins to appear at -200 mV when the passive layer starts to break down as the protective film is reduced. This corresponds to the dissolution of metal cations (primarily iron) into solution. These cations interact in a similar way with the sulphate anions as those cations produced during dissolution of pure iron. It is worth noting the striking similarity between the spectra in Figures 3(i) and 3(ii).

Such a similarity is most likely due to the same species produced during the anodic dissolution of iron and cathodic corrosion of SS304.

3.1 The Anodic Polarisation of Stainless Steel at Various pH Values

The SNIFTIRS spectra obtained during the anodic polarisation of stainless steel in sulphuric acid are displayed in Figure 4(i). The initial negative band at 1100 cm^{-1} corresponds to an increase in the concentration of free sulphate ions present in the double layer. The lack of any change in bisulphate bands at 1200 cm^{-1} and 1040 cm^{-1} would seem to exclude sulphate production from a bisulphate reaction due to an increase in pH. Another explanation could be the movement of the anions into the thin layer to balance the anodic shift as the potential is stepped from the open circuit potential (OPC) at -380 mV to 0 mV. In 1 mol dm^{-3} sulphuric acid solution the bisulphate anions will be in vast excess of the sulphate ion so these ions were expected to be predominant in any electrode processes. However, sulphate ions have a higher charge density than bisulphate and so would balance the potential change more efficiently.

As the potential is made more positive, this sulphate band appears to split into 3 separate bands at 1147 cm^{-1}, 1042 cm^{-1} and 980 cm^{-1}. These bands may be assigned to sulphate ion with a symmetry of C_{3V}, that is bonded to either the metal ions or surface of the electrode by a single co-ordination bond. This would correspond the formation of a unidentate, covalently-bonded complex. Therefore, the interaction between the sulphate and metal species is stronger during anodic polarisation than in the cathodic, reductive dissolution.

The anodic polarisation in neutral sulphate solution (shown in Figure 4(ii)) produces a single positive band at 1100 cm^{-1} corresponding to the sulphate ion vibration. This indicates a reduction in the amount of free sulphate in solution as the polarisation is increased anodically, repeating the trend seen in sulphuric acid. The appearance of three positive bands at 1190 cm^{-1}, 1040 cm^{-1} and 980 cm^{-1} again seems to indicate a reduction in the symmetry of the sulphate ion to C_{3V}. The higher wavenumber of the band near 1200 cm^{-1} may be due to an overlap of unidentate peak with that of bisulphate.

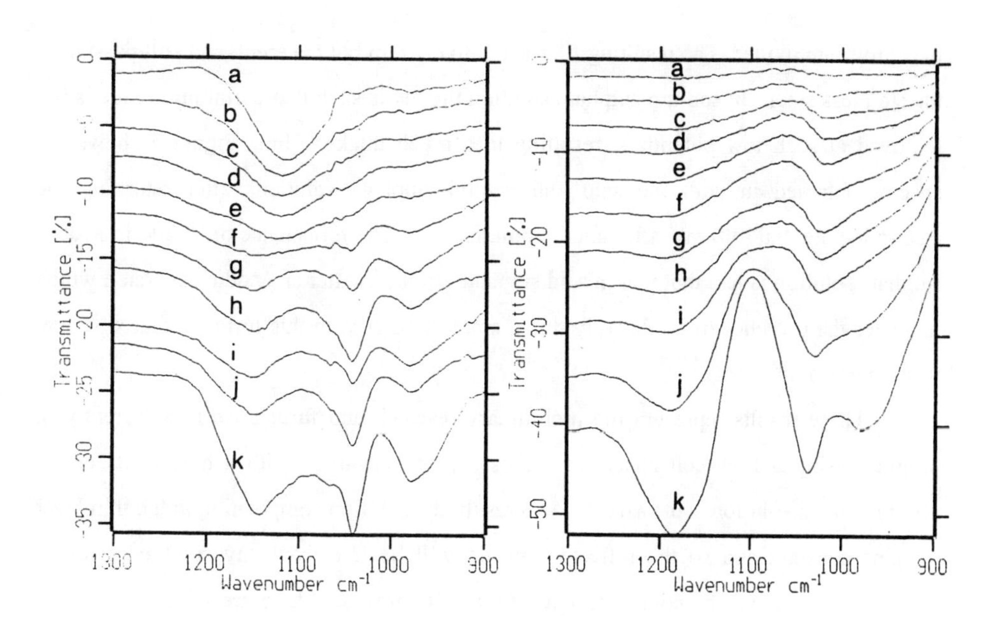

i) Anodic polarisation of stainless steel in 1 mol dm^{-3} *Sulphuric Acid. Spectra a, b, c... k correspond to* E_2 *= 0 mV to 1000 mV in 100 mV steps* E_1 *= -380 mV (OCP)*

ii) Anodic polarisation of stainless steel in 1 mol dm^{-3} *Sodium Sulphate Solution. Spectra a, b, c... k correspond to* E_2 *= 0 mV to 1000 mV in 100 mV steps* E_1 *= -50 mV (OCP)*

Figure 4 *SNIFTIRS spectra obtained during the anodic polarisation of Stainless steel (type 304)*

The assignment of these bands to surface or solution species cannot be made without further experiments. However, the magnitude of the absorption bands points towards solution species. Nevertheless, the formation of a film containing the anions on the electrode surface cannot be excluded. In fact, surface enhanced Raman spectroscopy (SERS) on iron at pH 5 in buffered sulphate solution has shown that bidentate sulphate was bonded into the film[8].

3.2 Film Formation after Anodic Polarisation

At low pH values (<3) part of the oxide film is removed by reductive dissolution of the chromium oxide[9]. The resulting film is relatively thin but coherent, and still passivates the stainless steel. In sodium sulphate solution the pH is such that chromium oxide is not removed at such low potentials, resulting in a much thicker film. Figure 5 shows the spectra obtained in both the acid and neutral sulphate solutions after returning the electrode potential to 0 mV after anodic polarisation. The reflectance of SS304 in neutral sulphate solution is less that that in acid sulphate solutions after an anodic excursion which indicates the presence of a relatively thick, amorphous film in this form compared to that in sulphuric acid.

These results represent the preliminary research and further work is currently in progress using a flow cell apparatus. This uses an electrode with a hole in its centre through which solution is drawn off. As a result, the solution composition in the thin layer remains constant and so the reflected spectra will be free from any soluble corrosion products. These will be compared with the results presented here are which are from a static cell and the spectra are made up of both the surface film and the thin solution layer.

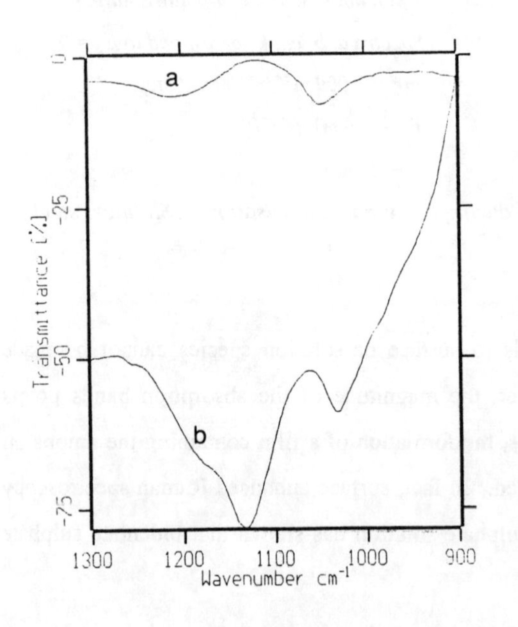

a) Film formed by anodic polarisation in 1 mol dm^{-3} Sulphuric Acid, E_2 = 0 mV after polarisation to 1000 mV. E_1=0 mV before polarisation.

b) Film formed by Anodic polarisation in 1 mol dm^{-3} (Sodium Sulphate Solution, E_2 = 0 mV after polarisation to 1300 mV. E_1=0 mV before polarisation.

Figure 5 *SNIFTIRS spectra of Stainless steel (type 304) after Anodic polarisation*

4. CONCLUSIONS

The results show that corrosion processes effecting stainless steels, such as the dissolution of metal ions and film formation, can be studied using the SNIFTIRS technique. The interaction of a solution species such as sulphate, that has well-established infrared characteristics, with the surface of a metal can help to elucidate the processes involved in its corrosion. The technique provides information not only the products of any reaction but also on the disappearance from the thin layer of reactants. Electrodes made from pure elements, which are components of SS304, such as chromium and nickel, could be used to help identify some of the products containing them.

The use of SNIFTIRS could be further improved by using a flow cell, which removes any spectral effects caused by species in solution, to solely examine the formation of the passive films formed on the electrode surface. At present, only work in the near IR range has been performed. In future, the far infrared region containing many of the important bands (particularly metal and metal-oxygen bonding) will also be studied.

Acknowledgements

The authors would like to thank EA Technology (Capenhurst) for the industrial studentship (JFH), MTD for the EPSRC studentship (LKWO) and the Royal Society of Chemistry for financial support.

References

1. A. Bewick, K. Kunimatsu and B. S. Pons, *Electrochim. Acta,* 1980, **25,** 465.
2. K. Nakamoto, 'Infrared and Raman Spectra of Inorganic and Co-ordination Compounds', Wiley, 1978
3. G. Okatmoto, *Corros. Sci..* 1973, **13,** 471.
4. B. S. W. Dawson, D. E. Irish and G. E. Toogood, *J. Phys.Chem.,* 1986, **90,** 334.
5. A. Bewick, M. Kalaji and G. Larramona, *.J. Electroanal.Chem.,* 1991, **318,** 207.

6. L.J. Obonsky and T.M. Devine, *Corros. Sci.,* 1995, **37,** 17.
7. H. Seki, *IBM J. Res. Develop.,* 1993, **37,** 227.
8. J. Gui and T. M. Devine, *Corros.Sci.,* 1994, **36,** 411.
9. M. S. El-Basiouny and S. Haruyama, *Corros.Sci.,* 1977, **17,** 405.

5. SCANNING ELECTROCHEMICAL MICROSCOPY (SECM): ITS APPLICATION TO THE STUDY OF LOCALISED CORROSON

Guy Denault, Lynn Andrews, Sarah Maguire and Steven Nugues

Department of Chemistry, University of Southampton, SO17 4BJ, UK

1. INTRODUCTION

Three dates, 1940, 1972 and 1986, mark the history of scanning electrochemical microscopy. The first report on the use of a scanning electrochemical probe was published in 1940 by Evans[1] who used a travelling reference electrode to map equipotential surfaces in solution as a means of calculating corrosion rates on a water line. The second significant article[2] was written in 1972 by Isaacs who improved Evans's approach by mounting a reference electrode on the arm of an x-y recorder and used this apparatus to produce two-dimensional maps of pitting events. He also coined the name 'scanning reference electrode technique' or SRET. The third most important article was published in 1986 by Bard[3] who described the principle of the scanning clectrochemical microscope, which was later given the acronym SECM. In this work a microelectrode mounted on the micropositioning stage of a scanning tunnelling microscope was used to map the topography of an electrode with micrometer spatial resolution. Before this study all scanning electrochemical probes had been potentiometric devices. The new apparatus was different because the probe was operated as an amperometric sensor and its faradaic current reflected the interaction between the sample surface and the probe. Since then the SECM has evolved significantly, it has been the object of over 100 publications and several reviews.

The aim of this article is to discuss the SECM and its applications in the context of localised corrosion. A brief review of the concepts and fundamentals of scanning electrochemical microscopy is presented. SECM theory is kept to a minimum sufficient for a proper understanding of the following sections; a comprehensive treaty of the theoretical background to SECM can be found in a recent review[4]. The next section considers a range of SECM applications related to corrosion studies, the discussion focuses on experimental approaches. Finally our conclusion aims to present a summary in the form of several instrumental configurations.

2. BRIEF REVIEW OF THE CONCEPTS AND FUNDAMENTALS OF SECM

The scanning electron microscope belongs to the family of scanning probe microscopies, where a sample is analysed by recording the response of a fine probe while rastering the tip of the probe at a very close distance above the sample surface. The SECM was invented by combining technological developments in scanning tunnelling microscopy with microelectrode applications. The sample (also called substrate) and the probe are part of an electrochemical cell and immersed in a solution containing an electrolyte and a redox mediator. The probe is held precisely in three dimensions by means of micropositioners. The principle of the technique is to control the electrochemistry of the microelectrode, approach the tip of the microelectrode very close to the sample surface and follow the interactions between the tip and the substrate by recording the microelectrode electrochemical response. Figure 1, below, illustrates the basic principle of the apparatus.

In its most common mode of operation, the feedback mode, the SECM is analogous to a high-resolution electrochemical radar[4] where a redox species electrochemically generated by the tip, diffuses away from the tip, interacts with the substrate and diffuses back towards the tip. The magnitude of the faradaic tip current reflects both the tip-substrate distance d and the reactivity of the sample surface below the tip. Hence the SECM has the ability to probe the topography and the reactivity of surfaces. Most studies reported so far have used this mode of operation and it has been thoroughly described in numerous publications.

The number of SECM studies is increasing, the range of applications is widening and new experimental procedures and new detection methods are being reported; it is therefore useful to consider the SECM in the general context of scanning electrochemical probes.

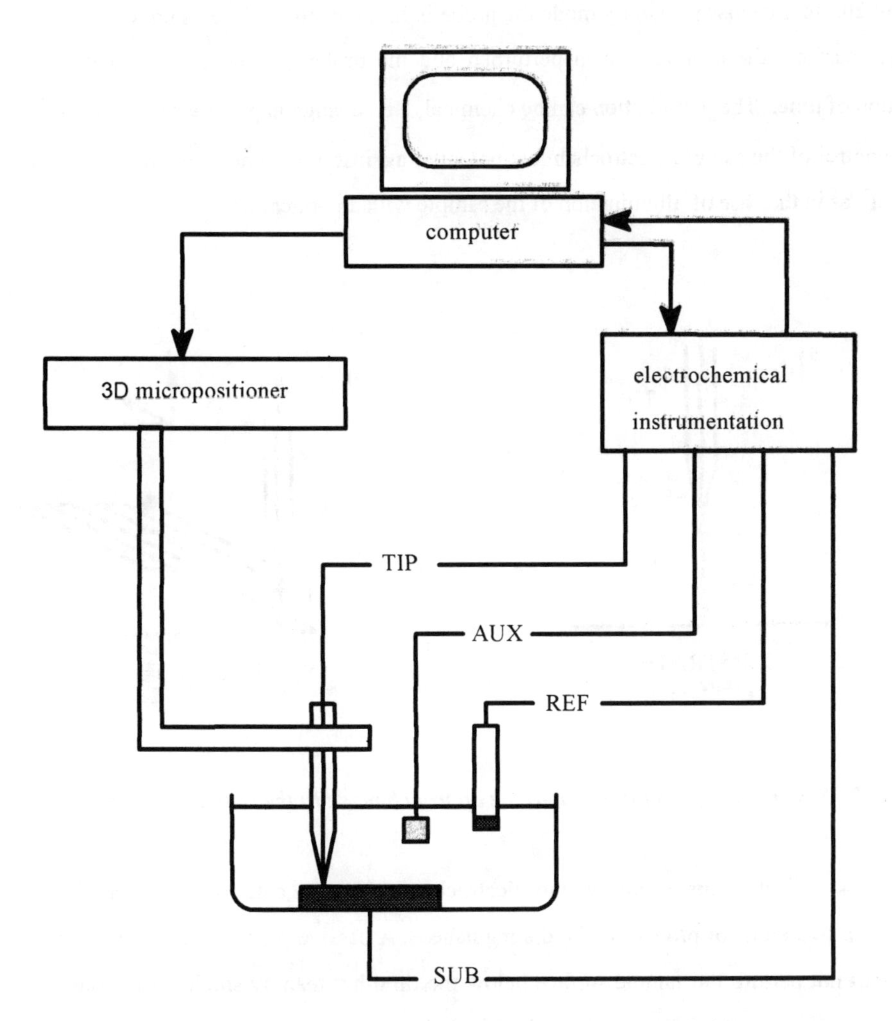

Figure 1 *Schematic representation of a scanning electrochemical microscope*

Three modes of operation can be distinguished depending on the movement of the tip with respect to the sample surface. In the approach mode, the probe response is recorded while moving the tip towards the surface, see Figure 2a, an approach curve is obtained by plotting the tip response against the tip-substrate distance, *d*. In the scanning mode, the probe is first moved close to the sample surface and the probe response is recorded while rastering the tip in a plane parallel to the sample surface. A map of the surface is obtained by plotting the tip response against the *x* and *y* coordinates, see

Figure 2b. In the close proximity mode the probe is first moved as close as possible to the sample surface, the sample is then perturbed and the probe response is recorded as a function of time. The perturbation can be chemical, e.g., change in pH, electrochemical[5,6], e.g., control of the sample electrochemical potential against a reference electrode, or even optical[7], as in the case of illumination of the sample with an optical fibre.

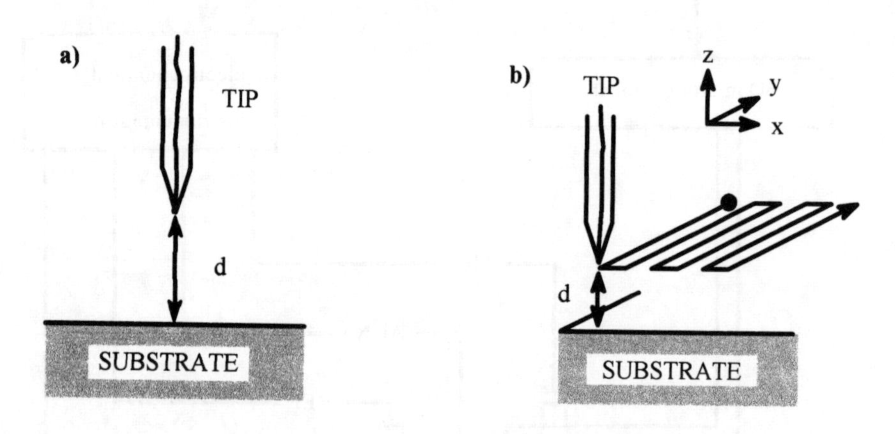

Figure 2 *Schematic representation of a) the approach mode b) the scanning mode.*

As indicated, the nature of the electrochemical probe determines the mode of detection; two kinds of probes can be distinguished. A passive probe is a microelectrode that does not perturb the sample surface below the tip other than by shielding it from the bulk solution, e.g., typical passive probes include micro reference electrodes and ion selective microelectrodes. In these two examples the tip is analogous to a potentiometric sensor and the SECM is said to operate in the potentiometric mode or potentiometric detection. With an ion specific microelectrode the tip response is related to the logarithm of the concentration of a redox species[5]. An active probe is a microelectrode, which either interacts directly with the substrate or with the solution. This interaction can take several forms; the probe may participate in an electrochemical reaction with the sample, generate a redox species that interacts with the sample, consume a redox species produced by the sample or collect a current generated by the sample or by another electrode. An active microelectrode carries a current and behaves as an amperometric sensor; the SECM is said to operate in the amperometric mode. In the most popular configuration, the feedback mode, the tip response is controlled by the tip potential and the tip current is proportional to the concentration of a redox species.

Besides the probe and its movement, it is necessary to consider the sample. It may be part of the electrochemical circuit or simply left unconnected. When the substrate is connected, as in Figure 1, its potential can be controlled by a potentiostat or bipotentiostat and the sample is said to be biased, alternatively the current flowing to the substrate can be controlled by a galvanostat. When the substrate is not connected, its potential is left under the control of the solution and is said to be unbiased. A schematic representation of the various configurations is presented elsewhere[8]. The amperometric mode is very versatile and leads to a large number of applications. One case in particular is worth mentioning for historical reasons. In the direct amperometric mode the sample and the probe are part of a two-electrode set and this arrangement was one of the first SECM experiments carried out (see Figure 3).

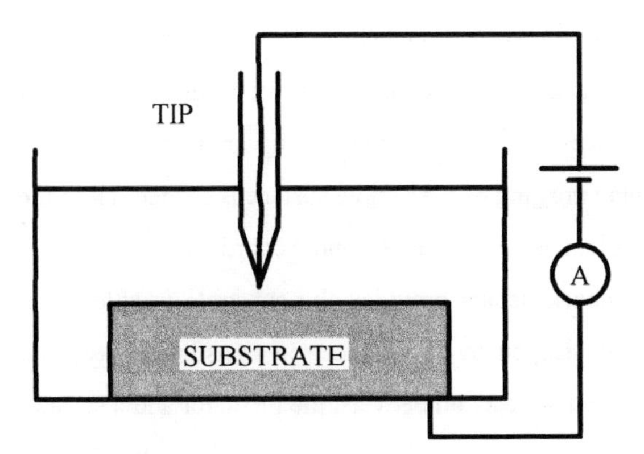

Figure 3 *Schematic representation of the direct amperometric mode.*

Scanning electrochemical microscopy has attracted much interest through the applications of the feedback mode previously described. In order to understand the principles, it is necessary to consider the amperometric response of a microelectrode under diffusion controlled conditions. Microelectrodes have unique properties which typically depend on a critical dimension (e.g. the radius for a microdisc electrode) when this dimension is smaller than 50 μm. For example, a microdisc leads to the very rapid establishment of a steady state current when the electrode potential is stepped from a value where no reaction occurs to one where a redox reaction is diffusion controlled. The diffusion layer is very small (of the order of 7 times the value of the radius of the microdisc)

The concept of SECM came from the observation that thc steady state current to the microelectrode is affected by the presence of the substrate when the tip-substrate distance is smaller than the diffusion layer. Far away in the bulk, the steady state tip current is governed by quasi-hemispherical diffusion current $i_{T,\infty}$ (for a microdisc, $i_{T,\infty}$ = $4nFDC_b a$, where C_b is the bulk concentration of redox mediator and the other parameters have their usual meaning). Close to an insulator, hindered diffusion leads to a tip current smaller than $i_{T,\infty}$; because of the blocking effect, $i_{T,\infty}$ tends towards zero as the tip substrate distance decreases; this is so-called negative feedback. Close to a conductor, feedback diffusion, where the mediator diffusing from the tip is regenerated by the substrate, leads to a tip current greater than $i_{T,\infty}$ increases exponentially as the tip approaches the substrate; this is known as positive feedback. For an illustration of these diffusion patterns, see Figure 2. Since the tip current depends on the tip-substrate distance, topographic imaging of the sample surface is carried out in the scanning mode by recording the tip current as a function of *x* and *y* coordinates.

Positive feedback occurs provided the substrate is able to regenerate the redox mediator. This may happen by direct electron transfer between the mediator and the substrate or via a chemical reaction between the mediator and a surface bound species. At a given tip-substrate distance, the ratio $i_{T,close}/ i_{T,\infty}$; indicates the rate of turnover of the redox mediator between the tip and the substrate. Any variation in the rate of regeneration affects the tip current. The tip has thus the ability to probe the local reactivity of the substrate. The situation is rather complicated since the magnitude of the tip current depends on the convolution between the tip-substrate distance and the surface reactivity. This is clearly observed in the approach mode where a range of surface reaction rates produces a family of curves limited by the two extreme cases corresponding to an insulating material (no regeneration) and a conducting material (infinitely fast regeneration). In the scanning mode, variations of tip current with tip-substrate distance reflect surface topography, while variations of tip current with the rate of regeneration at the substrate solution interface reflect surface reactivity. This characteristic is almost unique and the popularity of the SECM is in part due to its ability to map the reactivity of surfaces rather than their topography.

The spatial resolution is primarily determined by the size of the tip: the smaller the electroactive area of the tip, the smaller the steady state diffusion layer and the better the resolution. Numerical simulations have shown that the SECM could detect conducting islands as small as $a/20$ provided they were sufficiently far apart; the detection of insulating islands is much less favourable because hindrance diminishes with the dimension of the island. In other words, the magnitude of the tip current is much more sensitive to feedback diffusion than to hindered diffusion. Distinguishing surface reactivity from surface topography has been made possible by means of tip modulation with lock-in detection[9] and by convective enhancements of the tip current[10]. In both cases the ability to image the surface in the constant current mode yields an improvement in spatial resolution.

The discussion has so far focused on the operations of the SECM in the steady state; it is now worth considering the chronoamperometric response of the SECM, which forms the basis of several fundamental studies where the instrument is used as an analytical tool. Most of these applications are based on the close proximity mode, where the tip is brought very near the substrate and the tip response is monitored as a function of time following the application of a perturbation to the substrate. Since the electrochemistry of the tip and of the substrate can be controlled precisely, several SECM applications make use of the twin electrode geometry. For example, tip-substrate voltammetry yields two voltammograms: tip current versus substrate potential and substrate current versus substrate potential. The principle is analogous to experiments with the rotating ring disc electrode, the quartz crystal microbalance and the probe beam deflection method. SECM generation collection experiments lead to very high collection efficiencies due to the very high rates of mass transport achieved between the tip and the substrate. The steady state mass transfer coefficient of the SECM varies from D/a at infinite tip-substrate distance to D/d very close to the substrate. Experimentally, the tip can be moved to a distance equal to one tenth of the disc radius, thus providing a higher steady state mass transfer coefficient than that found with the same microdisc in the bulk solution.

3. SECM APPLICATIONS IN CORROSION STUDIES

Several publications have appeared which illustrate the applications of scanning electrochemical microscopy in the field of corrosion. White and co-workers[11,12] reported the results of a very interesting study of thin titanium oxide films. Using the Br^-/Br_2 redox mediator, they were able to image randomly located electroactive sites with characteristic diameters in the order of 50 μm. These sites proved to be precursors of the local breakdown of the oxide film and of the formation of pits in the titanium sample. Careful control of the titanium electrode potential was used to adjust the activity of the precursor sites which was reflected by the magnitude of the tip current peaks on the SECM image. In a subsequent article[13], they showed that the precursor sites could be imaged with other mediators thus demonstrating that these sites were not specifically related to the oxide/ Br^- chemistry but linked to inhomogeneities within the oxide film. In addition, localised activity was found to depend on the potential at which the redox species is reduced or oxidised and was observed with reactions occurring at potentials positive with respect to the conduction band edge of TiO_2.

Studies by Wipf[14] and by Gilbert[15] illustrate how the SECM can be used to detect and map metallic ions locally released by corroding sites. Gilbert also showed that the SECM is well suited to the detection and mapping of regions consuming oxygen naturally dissolved in the electrolyte.

Tanabe and co-workers reported an SECM study of pitting corrosion on austenitic stainless steels[16]; the article is interesting in many ways. Several new experimental approaches are described. Some of the images shown present one of the highest spatial resolution reported for SECM images (better than one micrometre). From the point of view of localised corrosion, the images are outstanding and to our knowledge, no other technique has produced such highly resolved images of the chemistry within actively corroding pits. For example, using a double potential step chronoamperometric approach they were able to assess, simultaneously, the local concentrations of H^+ and Cl^-. When performed during rastering, this technique yields two maps respectively showing the distribution of these ions inside an active pit. The article lacks a proper discussion of the results, e.g., it is not clear how much the magnitude of the current measured at very short times reflects the concentration of H^+ or Cl^- since the charging current must be rather large. Similarly, it would be useful to know how much of the current measured at positive tip potentials reflects the oxidation of Cl^- compared to the oxidation of Fe^{3+} species randomly located within the pit. Nevertheless, it illustrates the versatility and power of the SECM approach.

In our Group, the SECM has been used to map electrochemical activity above coated metal surfaces[17]. Figure 4 shows a tip current map recorded above a platinum electode coated with a Nuclepore membrane. The solution contained 10 mM $K_4Fe(CN)_6$ and 0.1 M KCl.. The substrate potential was held at +0.6 V vs. SCE to oxidise the redox mediator (ferrocyanide) to ferricyanide, while the tip potential was held at at -0.1 V vs. SCE to reduce ferricyanide. This image was therefore recorded in the substrate generation-tip collection mode. The membrane has well defined pores (10 µm diameter) which were used to model pin holes in a coating. Far away from a pore and in the bulk, the tip current is equal to zero because there is no ferricyanide to reduce. Close to a pore, however, the tip current becomes increasingly negative due to the reduction of ferricyanide ion produced by the substrate at the bottom of the pore. In Figure 4, dark, circular regions correspond to large negative currents associated with the presence of individual pores.

We have also been engaged in the development of SECM amperometric techniques for monitoring proton fluxes at the solid-liquid interface[18]. pH dependent tip reactions were chosen so that variations in the tip current reflected changes in the local pH around the probe, the diffusion controlled current for proton reduction had been shown to be an effective measure of the production and consumption of protons by the sample surface. The technique is very sensitive to the production of protons, even in very alkaline solutions. However, the consumption of protons can only be detected provided the background solution has a pH below 6. This problem is circumvented using oxide formation on the tip. The tip potential is held at the foot of the oxide formation wave;

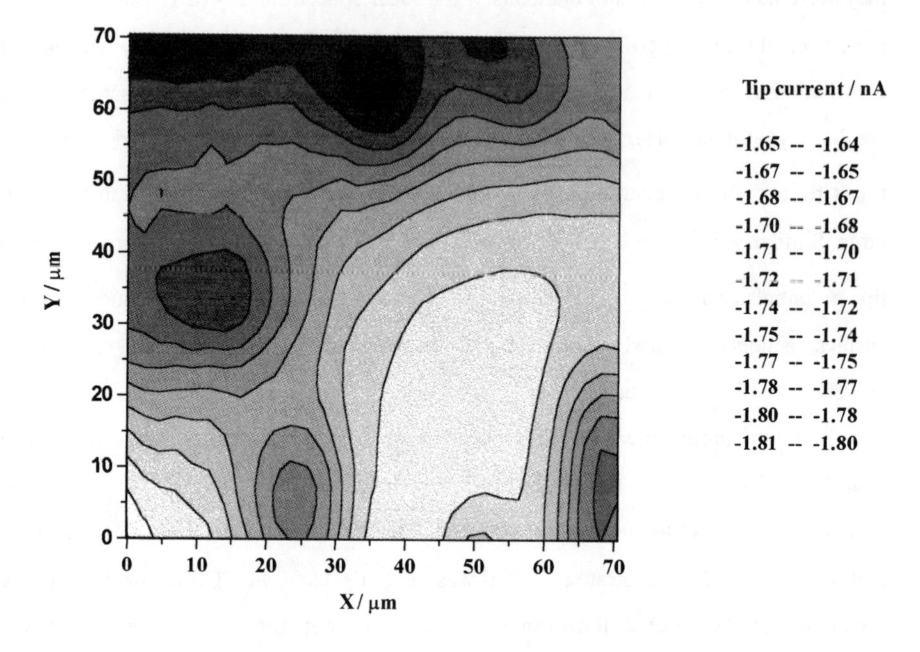

Figure 4 *SECM image recorded above a Pt substrate electrode coated with a 6 μm thick Nuclepore membrane (10 μm diameter pore) with a 2 μm diameter Pt microdisc in solution of 10 mM $K_4Fe(CN)_6$, 0.1 M KCl. The substrate potential was held at +0.6 V vs. SCE while the tip potential was held at -0.1 V vs. SCE.*

any variation in the local pH leads to the formation or reduction of tip oxide and respectively produces a positive or negative tip faradaic current. A similar approach is based on the current for oxygen evolution on the tip. In both cases the tip current is extremely sensitive to increases in the local pH. Experiments currently undertaken show that pH variations can be mapped above oxide covered surfaces even two hours after the formation of the oxide. Tip-substrate voltammetry and tip-substrate chronoamperometry were used to probe proton fluxes during hydrogen adsorption/desorption[19] and oxide formation/reduction[20] on platinum electrodes. In both cases, the substrate current yielded information about the transfer of electrons between the adsorbed layer and the metal surface while the tip current gave information about the transfer of protons between the adsorbed layer and the solution. Tip-substrate voltammograms showed that the transfer of electrons and protons are not necessarily simultaneous and were analysed in terms of various reaction mechanisms. Tip-substrate chronoamperograms, where the tip current is recorded following the application of potential step to the sample under study, provided an effective means of probing the transient nature of the proton fluxes during the surface reactions. Arrival of protons was detected within a few milliseconds following their release by the substrate, the magnitude of the tip current was found to depend strongly on the substrate potential and provided further evidence towards the reaction mechanisms proposed on the basis of the tip-substrate voltammograms.

CONCLUSIONS

The scanning electron microscope was shown to have an almost unique ability to probe (with micrometre spatial resolution) the distribution and magnitude of electrochemical activity on surfaces. Whether in the scanning mode or close proximity mode, the tip response is found to be highly sensitive to local variations of the concentrations of species (proton, chloride ions, dissolved oxygen, metallic ions) participating as reactant or product in the reactions at the surface of the sample under study. In this respect the SECM is well suited to the study of corrosion mechanisms and in particular to localised corrosion. The SECM is based on well-established electrochemical principles taken from microelectrode and thin layer cell theories. The instrument is very versatile, very flexible and several techniques, e.g. scanning reference electrode, can easily be implemented. With a minimum of modification the instrument can be turned into a whole range of scanning, electrochemical probes. The schematic diagrams shown in Figures 5-9 aim to summarise the various SECM experimental approaches, which have been reported and proposed to probe localised corrosion. In Figure 5, the tip probes the surface reactivity with the help of a redox mediator. In Figure 6, the tip generates a strong oxidising agent, which initiates pitting on the sample surface. In Figure 7, the tip probes amperometrically or potentiometrically, the local concentration of a species involved in the corrosion process. In Figure 8 the tip operates as a micro reference electrode as in the SRET and, in Figure 9, as a micro counter electrode used to directly map the local corrosion currents.

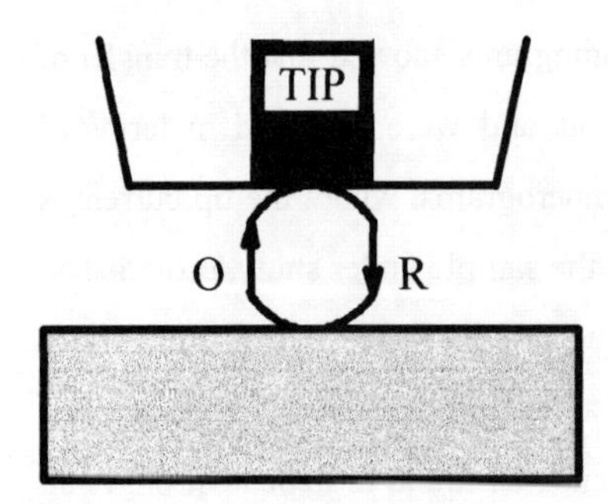

Figure 5: *SECM probing of surface electrochemical activity with a redox mediator.*

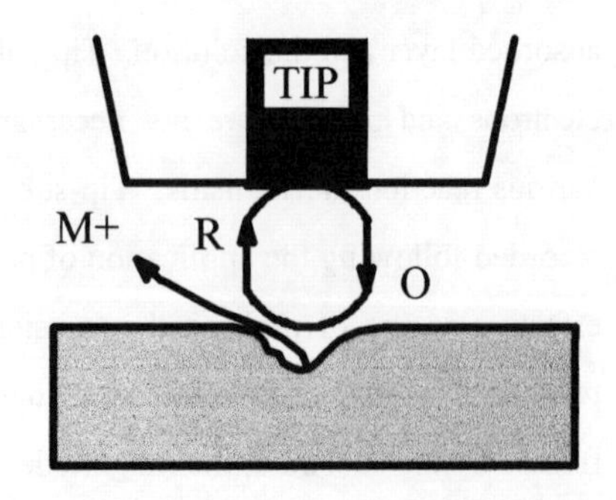

Figure 6: *Pit initiation by generation of an oxidising agent.*

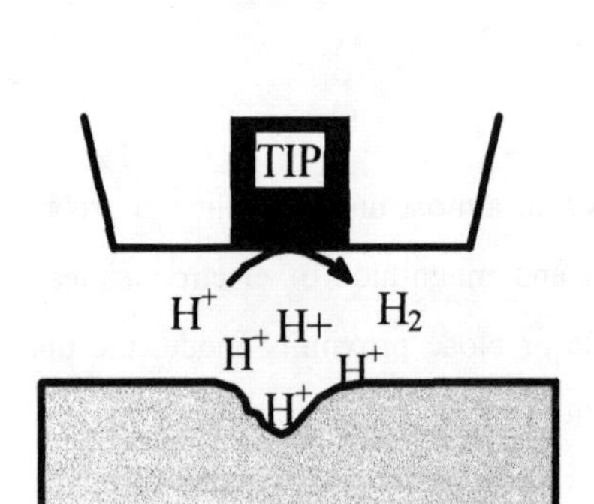

Figure 7: *SECM probing of local concentrations.*

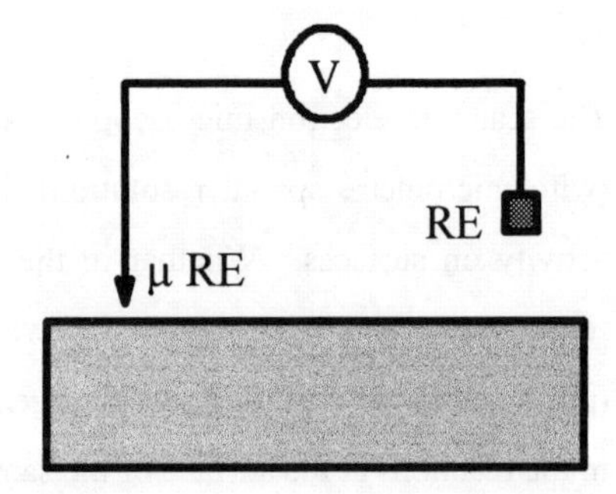

Figure 8: *Potentiometric mapping with a micro reference electrode.*

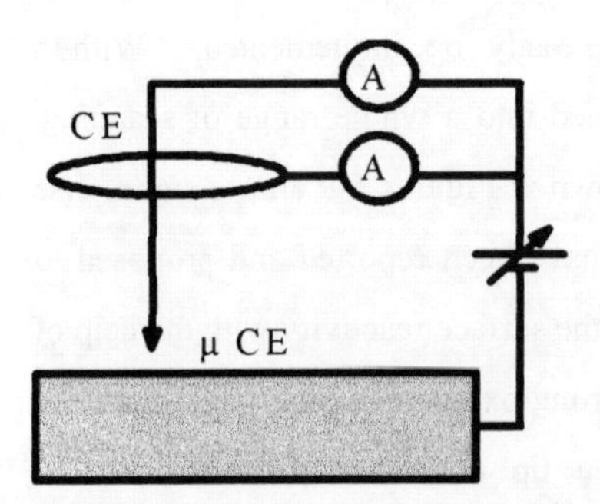

Figure 9: *Amperometric probing of localised corrosion currents with a micro counter electrode.*

Acknowledgements

The authors would like to acknowledge the following bodies for their financial support, EPSRC (Advanced Research Fellowship to Dr G.Denuault, grant GR/J01066) and a Unilever CASE award to Dr L.Andrews, Exxon Chemicals Ltd, the Southampton Electrochemistry Group, and Unilever Research, Port Sunlight.

References

1. U.R. Evans, *Journal of the Iron and Steel Institute,* 1940, **141,** 219.
2. H. S. Isaacs and G. Kissel, *J.Electrochem. Soc.,* 1972, **119,** 1628.
3. H-Y. Liu, F-R.F. Fan, C.W. Lin and A.J. Bard, J *Am. Chem. Soc.,* 1986, **108**, 3838.
4. A.J. Bard, F.F. Fan and M.V. Mirkin, in A.J.Bard (ed.), Scanning Electrochemical Microscopy in Electroanalytical Chemistry, Marcel Dekker, New York, 1994, **18,** 243.
6. G. Denuault, M. H. Troise-Frank and L.M. Peter, *Faraday Dis.,* 1992, **94,** 23.
7. M.H. Troise-Frank and G. Denuault, *J. Electroanal. Chem.,* 1994, **379,** 399.
8. N. Casillas, P. James and W.H.J. Smyrl, *J. Electrochem. Soc.,* 1995, **142,** L16-LI 8.
9. G. Denuault, M.H. Troise-Frank and S. Nugues, in A.A. Gewirth and H. Siegenthaler (eds.), Nanoscale Probes of the Solid/Liquid Interface, 1995, **Series E, 288,** 69-82. Dordrecht, Kluwer Academic Publishers.
10. D.O. Wipf, A.J. Bard and D.E. Tallman, *Anal. Chem.,* 1993, **65,** 1373.
11. K.Borgwarth, D. Ebling and J.Heinze, *Electrochim. Acta,* 1995, **40,** 1455.
12. N. Casillas, S.J. Charlebois, W.H. Smyrl and H.S. White, *J. Electrochem. Soc.,* 1993, **140,** L142.
13. N. Casillas, S.J. Charlebois, W.H. Smyrl and H.S. White, *J. Electrochem. Soc.,* 1994, **141,** 636.
14. S.B. Basame and H. S. White, *J. Phys. Chem.* 1995, **99,** 16430.
15. D.O. Wipf, *Colloids and Surfaces, A – Physicochemical and Engineeering Aspects,* 1994, **93,** 251.
16. J.L. Gilbert, S.M. Smith, E.P. Lautenschlager, *J. Biomed. Mat. Res.,* 1993, **27,** 1357.

17. H. Tanabe, Y. Yamamura and T. Misawa, *Mater. Sci. Forum,* 1995, **185-188**, 991.

18. S. Nugues, *PhD Thesis,* Southampton, 1996.

19. Yi-Fu Yang and G. Denuault, *J. Chem. Soc., Faraday Trans.*, 1996, **92,** in press.

19. Yi-Fu Yang and G. Denuault, *J. Electroanal. Chem.,* in press.

20. Yi-Fu Yang and G. Denuault, in preparation.

6. MARINE BIOFILMS ON STAINLESS STEELS: EFFECTS ON THE CORROSION BEHAVIOUR

D. Féron and I. Dupont

CEA-CEREM, Service de la Corrosion, d'Electrochimie et de Chimie des Fluides
B.P.6, 92 260 Fontenay-aux-Roses, France

1. INTRODUCTION

Stainless steels are widely used in marine environments, particularly in power plants and in offshore industries. Stainless steel surfaces exposed to natural seawater, are covered in a short time with an organic and biological layer, called a 'biofilm'. This biofilm settlement on these surfaces is often divided into at least three main periods: (i) an induction phase during which chemical adsorption of organic chemicals may occur on stainless steel surfaces and also the first adhesion of few bacteria, (ii) a second period corresponding to an increase in the number of adhesive micro-organisms, and (iii) a third period, a pseudo-stationary phase, during which the biofilm thickness is more or less constant (bacterial development and detachment). Adhesion of large marine organisms such as mussels, occurs during or after this last third period.

In the last decade, it has been shown by several laboratories that the open-circuit potential of stainless steels is shifted towards the noble direction for stainless steels exposed to natural seawaters during this biofilm formation and development[1-5]. This increase in the corrosion potential of stainless steels in natural seawaters is a major point in explaining the corrosion behaviour in seawater and mainly the localised corrosion which is promoted by potential ennoblement.

Our purpose is to focus on this potential ennoblement and consider the questions: does it appear at the beginning of the biofilm formation or when the biofilm settlement is achieved, what is the influence of the steel grade and the environmental conditions (seawater flow rate, temperature,)?

2. EXPERIMENTAL DETAILS

2.1 Materials

Eleven different stainless steel grades of European production were used during the reported tests. They include mainly high-grade materials: austenitic and duplex steels containing from 2% to 6% molybdenum (Table 1). Specimens included mainly tubes (diameter 23/25 mm, length 100 mm) and plates (60 x 60 x 3 mm or 200 x 300 x 3 mm).

Table 1 *Chemical composition of tested stainless steels (weight %, balance with Fe)*

Material	*Cr*	*Ni*	*Mo*	*N2*	*C*
UR B26	20	24.7	6.3	0.19	0.009
UR SB8	25	25	4.7	0.21	0.01
UR 47N	24.7	6.6	2.9	0.18	0.01
UR 52N+ (1)	25/25.7	6.3	3.6	0.25	0.01
654 SMO (1)	24.4/24.7	21.8/22.0	7.3/7.4	0.48	0.01
254 SMO	19.9	17.8	6.0	0.2	0.01
904L	19.8	24.2	4.3	0.051	0.01
316L (1)	17.2/17.3	11.1/12.6	2.5/2.6	0.05	0.017/0.035
SAF 2507	24.9	6.9	3.8	0.28	0.14
SAF 2205	22	5.5	3.2	0.17	0.15
SAN 28	26.7	30.3	3.4	0.07	0.019

(1) : several heats tested

Stainless steels were used as received. Before exposure, all the specimens were only degreased and then pickled at ambient temperature during 20 minutes in a solution containing 20% HNO_3 and 2% HF.

2.2 Exposure Conditions

All reported tests were performed at the CEA seawater facility, called the SIRIUS facility, which is located at La Hague (Normandy, France). The seawater was pumped at high tide on the rocky coast. The volume of the reservoirs was 120 m^3 maximum. There was no light between the pumping system in the sea and the exposed coupons. In the SIRIUS facility, the seawater was heated by a series of heat exchangers and was completely renewed via a single pass loop, no recirculation of the seawater occured.

The seawater flowed at 1.5 m^3 h^{-1} inside the stainless steel tubes (1 m s^{-1}) and inside the titanium autoclaves (0.01 m s^{-1}). When titanium autoclaves werc used, seawater was heated to 40°C in 15 minutes, but when there was no autoclaves, the time to heat seawater to 40°C was less than two minutes.

2.3 Measurements

The free corrosion potential of tubes and plates was measured versus the saturated calomel electrode (SCE) every hour by means of a Solartron Schlumberger data acquisition system (3351 D, ORION). Galvanic currents were also recorded every hour with the same device: the measurements are performed using an external resistance of 10 ohms for each galvanic couple (tubes or plates). Seawater temperatures, pH, redox potentials (platinum potential versus SCE) were also recorded every hour. Chemical analyses of the seawater were also performed during the tests: they include the main cations and anions, dissolved oxygen, total conductivity, organic matter content and chlorophyll a. Mean values obtained during the reported tests are shown on Table 2: no major variations were observed, except the ambient temperatures of the seawater which varied from 7°C in winter up to 17 °C in summer.

Table 2 *Seawater main composition during the reported tests (mean values with standard deviation)*

Conduct. mS cm^{-1}	pH	Cl^- g kg^{-1}	SO_4^{2-} g kg^{-1}	Na^+ g kg^{-1}	K^+ g kg^{-1}	Mg^{2+} g kg^{-1}	Ca^{2+} g kg^{-1}	O_2 mg kg^{-1}	Organic mg kg^{-1}	Chloro a. μg kg^{-1}
40.0 (2.5)	8.0 (0.1)	18.0 (1.2)	2.7 (0.1)	10.8 (0.2)	0.5 (0.1)	1.7 (0.2)	0.7 (0.1)	7.9 (0.7)	7.0 (3)	0.3 (0.2)

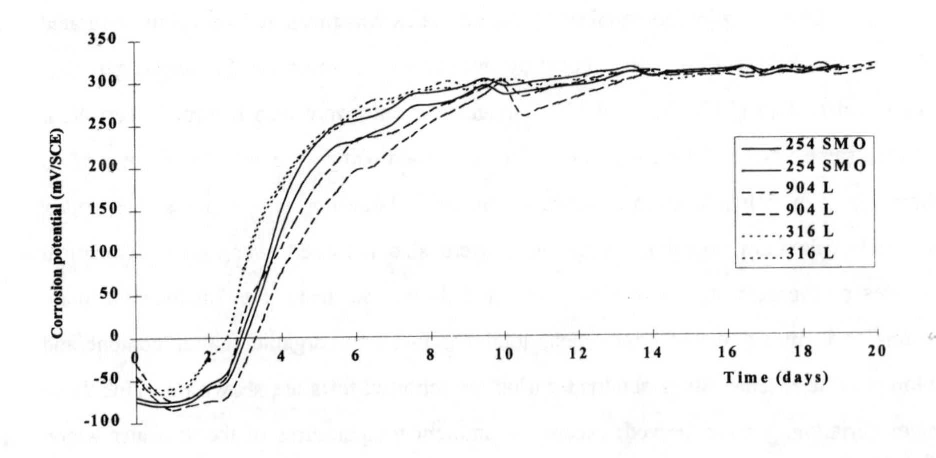

Figure 1 *Evolution of the corrosion potentials of stainless steels in natural seawater at ambient temperature*

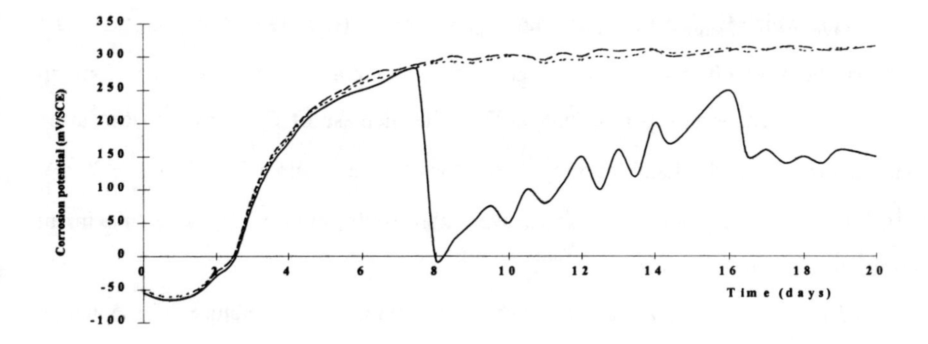

Figure 2 *Evolution of the corrosion potentials of 316L plates in natural seawater (dash lines : no corrosion on 316L coupons, continuous line : crevice corrosion on the coupon)*

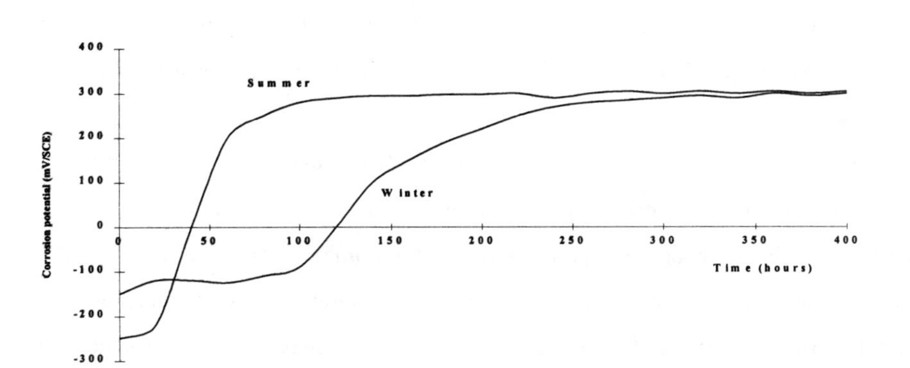

Figure 3 *Season influence on the evolution of the corrosion potential of stainless steels (654 SMO) plates in natural seawater at ambient temperature*

3. RESULTS AND DISCUSSION

3.1 Free Corrosion Potential Measurements

Figure 1 illustrates the evolution of the free corrosion potentials (E_{corr}) of stainless steel coupons when they are exposed to natural seawater at ambient temperature. Three main periods are observed whatever is the season or the grade of stainless steel. A first period during which E_{corr} is nearly constant at a low value (-100 to -300 mV *vs.* SCE) and which lasts between 40 to 150 hours in our experimental conditions. This first period will be called the «incubation » period below. Then the increase of E_{corr} takes place during about 100 hours. The third and last period corresponds to a constant E_{corr} at 300 ± 50 mV *vs.* SCE with relatively small variations, even after 3000 hours of exposure (maximum exposure time of our experiments).

3.1.1 Effect of stainless steel grade. At ambient temperature, the potential ennoblement occurs with all tested stainless steel grades: 316L, 904L, URANUS B26, SB8, 47N and 52N, SAF 2507 and 2205, AVESTA 654 SMO and 254 SMO, SAN 28. No differences are observed in terms of time to reach the maximum potential or values of this maximum potential. This result is illustrated in Figure 1 where the E_{corr} evolution of 316L, 904L and 254 SMO are compared in natural seawater at ambient temperature. If corrosion occurs during the exposure, the potential of the corroded coupons does not increase as described above and may decrease as illustrated in Figure 2 where a 316L coupon suffers crevice corrosion after 7 to 8 days of exposure to natural flowing seawater at ambient temperature.

3.1.2 Seasonal influence. Exposure of 654 SMO tubes and plates have been performed in spring, summer, autumn and winter. As shown in Figure 3, the incubation period varies from 40 hours in summer (mean seawater ambient temperature : 17.4°C) to some 110 hours in winter (mean seawater ambient temperature : 8.7°C) for stainless steel plates. In spring and autumn, the incubation time is between these two extreme values. But the maximum E_{corr} values are the same whatever is the season, about 300 mV *vs.* SCE under our experimental conditions.

3.1.3 Coupon influence. Two types of stainless steel coupons were used, as previously reported: plates and tubes. The same evolution and the same maximum values of E_{corr} were obtained on plates and tubes. But the incubation times were always longer for tubes (some 10 hours following the exposure conditions, mainly the season). This difference is probably due to the seawater flow which was low on plates (about 0.01 m s^{-1}) and higher in tubes (1 m s^{-1}).

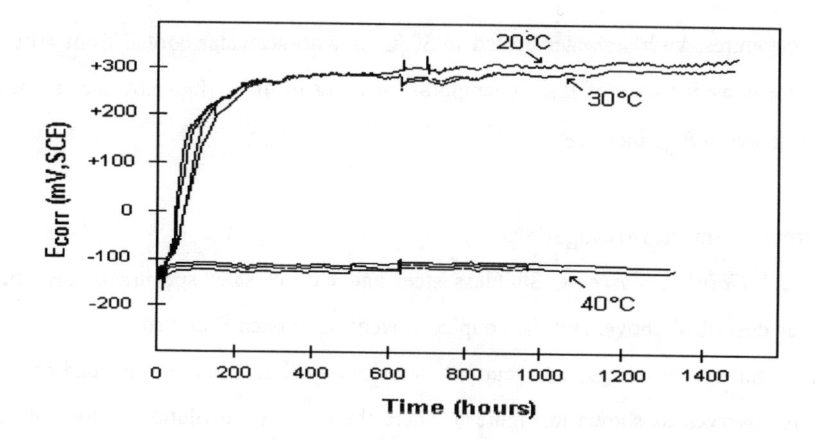

Figure 4 *Free corrosion potential evolution of stainless steel (654 SMO) tubes in flowing natural seawater (1 m.s^{-1}) at 20°C, 30°C and 40°C*

3.1.4 Effect of temperature. Experiments were conducted at the SIRIUS facility, at three temperatures, with twenty 654 SMO tubes at each temperature. Free corrosion potential evolutions are illustrated by the Figure 4 for these tubes and may be summarised as follows. At 20°C and 30°C, previous results obtained at ambient temperature are confirmed with a rapid increase in the corrosion potentials of the stainless steel tubes up to +300 mV *vs.* SCE in about 10 to 12 days. At 40°C, the free corrosion potential of all the tested stainless steel tubes remains constant at about -100 mV *vs.* SCE which is the starting value of the potentials at 20°C and 30°C.

Electrode potential ennoblement occurs also on platinum and titanium. At 20°C, titanium E_{corr} values reach ≈250 mV *vs.* SCE after 30 days of exposure, at 30°C, they reach ≈150 mV *vs.* SCE but at 40°C, titanium E_{corr} is nearly constant (≈10 mV *vs.* SCE after 30 days of exposure). On platinum, high potential values (270 to 300 mV *vs.* SCE) are obtained whatever the temperature is, while the starting values of the redox potentials are nearly the same at 20, 30, or 40°C: between 100 and 200 mV *vs.* SCE.

In another series of tests, five superaustenitic tubes are exposed to seawater heated at 30°C and five other tubes are located in the section where the same seawater is cooled at 26°C after being heated up to 40°C. The E_{corr} evolution of these tubes is the following: at both temperatures, with seawater heated to 30°C or with seawater cooled from 40°C to 25°C, there is an increase of E_{cor}. A temporary heating to 40°C does not alter seawater "enough" to avoid E_{corr} increase.

3.2 Electrochemical Investigations

3.2.1 Galvanic currents. Stainless steel and carbon steel specimens have been coupled as described above, and the coupled current have been followed. In all cases, an important increase in the galvanic current (stainless steel is cathodic and carbon steel anodic) is observed as shown in Figure 5 where the observed evolution of the galvanic current is reported for three seasons : winter, spring and summer. Figure 6 shows that the galvanic current increase after a longer incubation time in winter (350 hours) than in summer (120 hours), in accordance with the E_{corr} evolutions. It could mean that the cathodic properties of the stainless steel are affected by the biofilm formation at ambient temperatures, in natural seawater.

3.2.2 Polarisation currents. Stainless steel plates (AVESTA 254 SMO) have been polarised at plus and minus 20 mV around E_{cor}, during their exposure to natural seawaters at ambient temperature. The polarisation current is the current which is needed to polarise the specimen of ± 20 mV *vs.* E_{corr} Before the E_{corr} increase, the polarisation currents are constant between 5 and 10 $\mu A\ cm^{-2}$ (E_{corr} between -200 and -150 mV *vs.* SCE). The polarisation currents reach 20 to 40 $\mu A\ cm^{-2}$ when the stainless steel ennoblement takes place (E_{corr} increases to +250 mV vs. SCE). This increase in the polarisation current which takes place with an increase in the corrosion potential of the stainless steel, means that it is the cathodic reaction (oxygen reduction in the case of aerated seawater) on stainless steels which is faster or which is replaced by a faster cathodic reaction when biofilm formation occurs at ambient temperatures in natural seawater.

3.2.3 Oxygen influence. The influence of oxygen has been checked during one experiment where deaerated seawater was used at ambient temperature. The stainless steels E_{corr} were found to be constant between -100 and 0 mV vs.SCE during the tests as shown in Figure 6: no stainless steel ennoblement occurs without oxygen. This result points out the importance of oxygen in the ennoblement process.

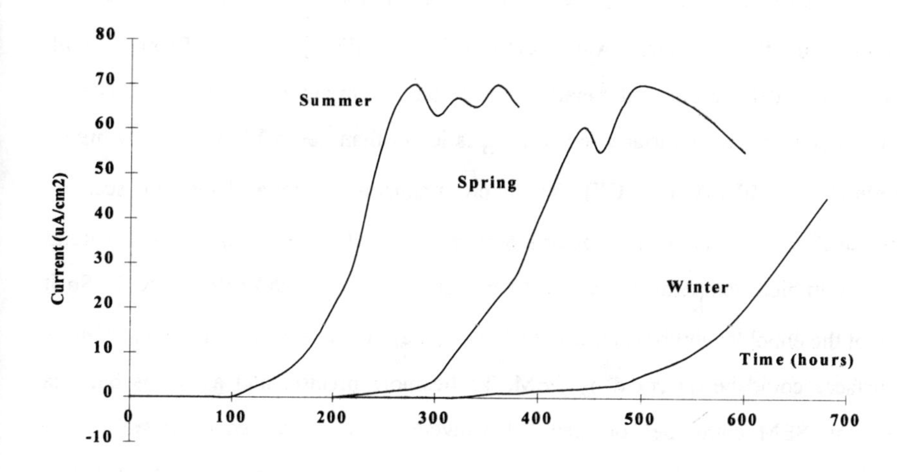

Figure 5 *Evolution of the current between stainless steel (cathodic) and carbon steel (anodic) exposed to natural seawater (coupling current in μA cm^{-2}, anodic and cathodic areas are the same)*

3.3 Biofilm Settlement

3.3.1 Masses of biofilm. Sampling of biofilms were performed on exposed coupons and the first investigations were performed on the biofilm masses. In fact, the biofilm masses were always very low at the beginning of the seawater exposure: less than 0.1 mg cm^{-2}, when the free corrosion potentials of the stainless steel coupons are lower than 200 mV *vs.* SCE, on coupons exposed at ambient temperature. Generally speaking, biofilm masses increased with exposure time, so with E_{corr} at the beginning of exposure. But, as shown in Table 3 where are reported the biofilm masses collected on tubes exposed in natural flowing seawater at 20, 30 and 40°C, biofilms are present on all the stainless steel tubes, including studies at 40°C. The fact that there is no E_{corr} increase at 40°C, in spite of a massive presence of biofilm, clearly demonstrates that biofilm settlement is possible in seawater without ennoblement of stainless steel.

3.3.2 SEM observations. Biofilm observations were performed with a Scanning Electron Microscope on stainless steel coupons, at different values of E_{corr}. A special preparation of the samples was performed, including biofilm fixation with glutarhaldehyde, dehydration and metallization. In fact, no microorganisms were observed by SEM on stainless steel surfaces when E_{corr} is lower than 200 mV vs. SCE. At higher potentials (E_{corr}>250 mV *vs.* SCE), few microorganisms (bacteria shape) are seen on stainless steel coupons: about 10^5 bacteria per cm^2 after 360 hours of exposure in natural seawater at ambient temperature during the winter test (E_{corr} = 253 mV *vs.* SCE). So it seems that the ennoblement of stainless steel occurs before bacteria settlement on stainless steel surfaces could be observed by SEM. To be more precise, and as far as bacteria observed by SEM could be considered as irreversibly adhesive (ennoblement occurs before the irreversible adhesion of bacteria on the surfaces under our experimental conditions).

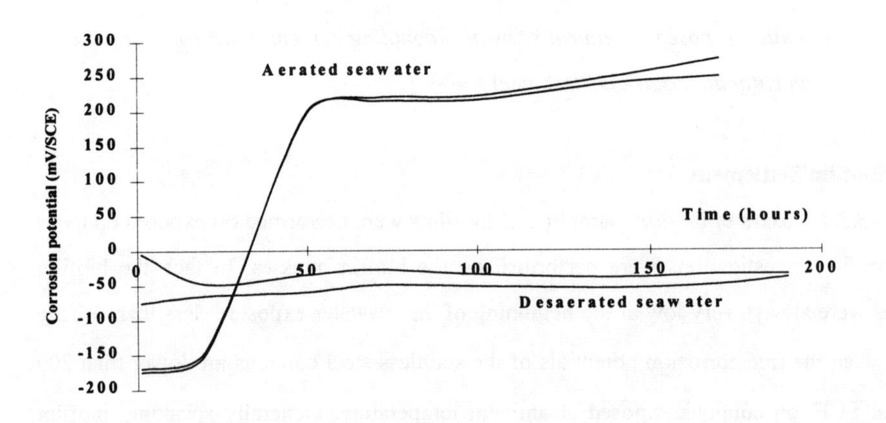

Figure 6 *Evolution of stainless steel (316L) potentials in aerated and deaerated seawater*

Table 3 *Biofilm settlement on stainless steel surfaces exposed to natural seawater (Epifluorescence observations of the percentages of the stainless steel surface covered with bacteria)*

Exposure time /hour	20°C		40°C	
	Potential *vs.* SCE/mV	% of surface covered	Potential *vs.* SCE/mV	% of surface covered
24	-110	5	-120	4
48	0	11	-135	20
72	200	10	-115	40
96	250	20	-	-
144	-	-	-140	40
240	290	25	-	-
312	-	-	-140	41

Table 4 *Biofilm sampling on stainless steel tubes exposed to seawater thermally altered*

Temperature	20°C		30°C		40°C	
Duration	7 days	27 days	7 days	27 days	15 days	27 days
Potential *vs.* SCE/mV	25	315	200	300	-90	-120
Wet biofilm content /mg cm^{-2}	0.22	14	0.32	11	0.74	4.4

3.3.3 Direct epifluorescence observations. More information on the biofilm settlement have been obtained by direct epifluorescence on exposed stainless steel surfaces. This technique allows one to count the total number of bacteria which colonise a surface, whatever is the adhesion of bacteria on this surface, reversible or irreversible. During experiments performed on the SIRIUS facility, the direct observations of microorganisms on surfaces by epifluorescence show that bacteria are gathered and numerous on stainless steel surfaces, and so difficult to count, even at the beginning of exposure. In a first attempt, the biofilm settlement on the surfaces is evaluated by the percentage of the surface covered by microorganisms (dead or alive) and eventually by substances which react also with the fluorescent substance (DAPI). These observations are performed at different exposure times and at different E_{corr} values. The results obtained on coupons exposed at 20°C and at 40°C are summarized in Table 4. They show clearly that the percentage of the surfaces covered by microorganisms is the same or even more important at 40°C than at 20°C whereas the free corrosion potential at 40°C stays constant and increases at 20°C. It shows clearly that the presence of a biofilm and bacteria on surfaces does not involve by itself the potential ennoblement.

3.3.4 Microbiological investigations. Investigations are then performed on the number of bacteria which are present in the biofilm to determine if there is a major difference between the bacterial colonization at 20°C and at 40°C which may explain the electrochemical behaviours. Aerobic, facultative anaerobic and strict anaerobic bacteria are isolated and counted. As shown in Table 5, there is no major difference between the number of bacterial populations, whatever the counting method.

The potential ennoblement observed on stainless steel surfaces at 20°C could not be explained only by the presence of bacteria on stainless steel surfaces: there is no differences concerning the global parameters of surface colonization (mass of biofilm, percentage of covered surface by the biofilm, total number of bacteria or number of aerobic, anaerobic or facultative bacteria) between 20°C and 40°C; the potential of stainless steel becomes more positive at 20°C but not at 40°C.

Table 5 *Number of bacteria on stainless steel surfaces (200 hours of seawater exposure)*

Exposure Temperature	Free corrosion potential	Total number of bacteria	Marine Heterotroph Bacteria (bacteria/cm^2)				
			Aerobic		Facultative		Anaerobic
	vs. SCE/mV	(1)	(2)	(3)	(2)	(3)	(2)
20°C	+302	4.4×10^7	6.0×10^4	2.1×10^4	2.3×10^1	1.4	<0.75
40°C	-133	3.3×10^7	1.9×10^5	7.4×10^4	0.9	2.5×10^1	2.3

(1) by epifluorescence (number of bacteria/cm^2), (2) by MPN method, (3) by plate count

4. CONCLUSIONS

The noble shift in the free corrosion potential of stainless steels in natural seawater occurs at ambient temperatures, 20°C and 30°C under our experimental conditions and is observed for all the other tested stainless steels: 654 SMO, 254 SMO, SAF 2507 and 2205, SAN 28, URANUS B26, SB8, 47N and 52N, 316L and 904L. The corrosion potentials rise from about -200/-100 mV *vs.* SCE to some +300 ± 50 mV *vs.* SCE. At ambient temperatures, stainless steel ennoblement is promoted in summer, but it occurs also in winter. At 40°C, the free corrosion potentials of stainless steels are constant at about -100 mV *vs.* SCE while biofilm formation and bacterial settlement on stainless steel surfaces occur at the three tested temperatures.

Biofilms developed at the three tested temperatures do not present differences in term of global parameters (masses, bacterial numbers, aerobic and anaerobic bacteria, etc.). Future work will concern the metabolic or the enzymatic activities of the bacteria. The chemical compounds (hydrogen peroxide and organic acids for instance) generated by the metabolic reaction of bacteria may be the key parameters for an explanation for the positive shift in the free corrosion potential observed at 20°C and 30°C.

Acknowledgements

This work was supported, in part, by the Commission of the European Communities under MAST and STRIDE contracts.

References

1. J.P. Audouard *et al.*, Proceedings of the Third EFC WP Microbial Corrosion, Estoril (Portugal), 1995, Paper No. 15 .
2. J.P. Audouard *et al.*, Proceedings of the International Congress on Microbial Induced Corrosion, New Orleans (Louisina-USA), 1995, Paper No. 3.
3. A.Mollica and A.Trevis, 4th Int. Congress on Marine Corrosion and Fouling, 1976, Juan-les-Pins, France.
4. E. Bardal, J.M.Drugli and P.O.Gartland, , *Corrosion Science*, 1995, **35**,(1-4), 257.
5. P. Chandrasekaran and S.C. Dexter, Corrosion'93 NACE conference, 1993, Paper No. 493.

7. THE INFLUENCE OF CHLORIDE IONS ON THE CORROSION PERFORMANCE OF DHP-COPPER AND 90/10 COPPER-NICKEL-IRON

D. Wagner[1], H. Peinemann[2], H. Siedlarek[3]

[1] *Institut für Instandhaltung GmbH, Kalkofen 4, 58638 lserlohn, Germany Technical*
[2] *Märkische Fachhochschule, Laboratory of Corrosion Protection, Frauenstuhlweg 31, 58644 Iserlohn, Germany*
[3] *Prymetall GmbH & Co KG, Zweifaller Strasse 130, 52224 Stolberg, Germany*

Abstract

DHP-copper and 90/10 copper-nickel-iron show a repassivating pitting attack in chloride ion containing electrolytes in concentrations typical for potable water. With increasing chloride ion concentration a change in the corrosion mechanisms of both materials was observed as indicated by a drastic increase of the mass loss during anodic polarisation. These observations were investigated in detail performing potentiostatic series and cyclovoltammetric experiments accompanied by relevant analytical and spectroscopic techniques to evaluate the change in the corrosion reactions.

1. INTRODUCTION

Pure copper has not been used extensively in marine applications since the days of copper bottomed, wooden hull sailing ships[1]. However, the metal provides the basic suite of properties from which modern marine copper alloys are derived. These properties are, among others, the general corrosion resistance to seawater, resistance to stress corrosion cracking and crevice corrosion as well as marine biofouling, excellent ductility and fabrication characteristics and high thermal conductivity. The various alloying additives guarantee additional strength and corrosion fatigue resistance, resistance to erosion and impingement attack, improved resistance to corrosion by both clean and polluted seawater and suitable casting and wrought fabrication characteristics [1].

Copper nickel alloys represent one of the most important copper alloy groups for marine applications. They are single phase alloys containing specific additions of iron and manganese which enhance their corrosion resistance in seawater due to the growth of a protective corrosion product film [1,2]. The formation of reaction layers of copper alloys in chloride ion containing aqueous environments is considerably influenced by the concentrations of the chloride ions. In seawater, the corrosion product formed on 90-10 copper nickel is predominantly a thin tightly adherent layer of cuprous oxide underneath a loosely adherent porous layered structure consisting primarily of cupric hydroxy chloride[3-7]. The morphology, composition and growth kinetics of the protective film on CuNi alloys have been described extensively in literature [3,4,8-11]. Experimental results of the same authors show that alloying additions of nickel and iron to copper improve corrosion resistance. The mechanism is proposed as the incorporation of nickel(II) and iron(III) ions into the highly defective, p-type cuprous oxide corrosion product film as "dopants" altering the defect structure [3,4,8-11]. The relevant manifestation of corrosion of 90-10 copper nickel in seawater is general attack [12].

Chloride ions in concentrations typical for potable water will initiate pitting in copper [13-15]. A two-fold reaction layer is formed on a copper surface, copper(I)-chloride underneath copper(I)-oxide. This amorphous copper(I)-oxlde is formed via hydrolysis of copper(I)-chloride and inhibits anodic metal dissolution [16,17]. The observed manifestation of corrosion is repassivating pitting. However, when the pH stays below 3.8 within the pit, repassivation is inhibited [15,17].

These observations indicate a different corrosion behaviour of copper alloys in chloride ion containing electrolytes with various concentrations. This contribution deals with the evaluation of the concentration influence on the corrosion performance of DHP-copper and 90/10 copper-nickel-iron. Potentiostatic series are most suitable to answer this question [18]. Quasi-stationary potentiodynamic experiments provide further information.

2. EXPERIMENTAL DETAILS

Potentiodynamic and potentiostatic experiments were performed in a Faraday cage in the absence of light, because light has a drastic influence onto the corrosion reactions in the system copper/water [15,19]. Working electrodes were prepared from hard copper tubes (DHP-Cu) and 90/10 copper-nickel-iron with a diameter of 22 mm. Rings with an outer surface of 8 cm^2 were pretreated with emery paper, polished with diamond paste and electropolished in orthophosphoric acid (70 % w/w) for 30 s with an anodic current density of $j = 0.2$ A cm^{-2}. The obtained rings were positioned in an electrode holder constructed in a way that only the outer surfaces of these rings were polarised using a conunon three electrode arrangement [17]. A ring of a platinum wire was positioned concentrically around the working electrode and was used as a counter electrode. A calornel electrode ($Hg/Hg_2Cl_2/KCl$ sat.) served as a reference electrode. Electrode potentials were recalculated versus the standard hydrogen electrode (SHE) in this paper.

A Haber-Luggin capillary was used to diminish the ohmic voltage drop. The electrode was stirred by a magnetic stirrer, and temperature of the electrolyte was kept constant at 20 °C. Sodium chloride electrolytes were prepared with double distilled water and chemicals of p.a. grade, The electrolyte within the cell was exchanged at a rate of 1.5 dm^3 d^{-1}. It was aerated with laboratory air. Carbon dioxide was removed by passing the laboratory air through a concentrated sodium hydroxide solution.

In case of solutions with small concentrations, ohmic voltage drops could not be completely avoided. The ohmic voltage drop was evaluated from potential transients measured with an oscilloscope (HAMEG, HM 408) after cut-off of the potentiostatic experiments at the end of immersion. Electrode potentials shown in the figures correspond to the off potentials. Potentiodynamic experiments were performed using a set-up with automatic IR compensation (e.g. EG & G, Princeton Applied Research, Model 273 A).

After finishing the potentiostatic series copper electrodes were pickled in 10 % w/w citric acid to remove the corrosion products. Then the change in surface contour was determined according to DIN 50 905[20]. Pit depths were measured using a stero microscope by focussing the non-attacked surface and the bottom of the deepest pit.

The experimental details of the potentiostatic and potentiodynamic experiments are summarised in Table 1.

Table 1 *Details of Potentiostatic and Potentiodynamic Experiments*

Working Electrode (WE)	*DHP-Cu* *CuNi10Fe*
Electrode Area	Outer surface: 8 cm^2
Pretreatment	Polished, electropolished, potentiostatic series: prepolarised (10 min at -800 mV_{NHE})
Counter Electrode (CE)	Platinum wire (0.5 mm)
Reference Electrode (RE)	Hg/Hg_2Cl_2/KCI sat. (242 mV_{NHE})
Electrolyte	NaCl
Electrolyte Concentration	200 mg dm^3 ÷ 20.000 mg dm^3 Cl^-
Exchange Rate Electrolyte	1.5 dm^3 d^{-1}
Aeration	Laboratory air
Duration	Potentiostatic series: 140 h
Scan rate	Potentiodynamic series: 0.01 mVs^{-1}

3. RESULTS AND DISCUSSION

Figure 1 shows the free corrosion potentials obtained for DIP-Cu and CuNi10Fe thirty minutes after immersion into the electrolyte as a function of chloride concentration. With increasing chloride concentration the free corrosion potential decreases with $\Delta E_{Fcorr} / \Delta c(Cl^-_{aq})$ = 85 mV per decade chloride concentration. No significant difference between the two materials was detectable.

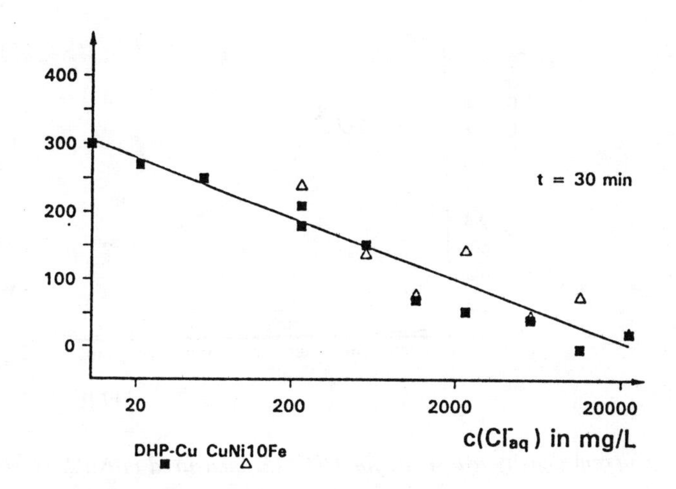

Figure 1 *Free corrosion potentials of DHP-Cu and CuNi10Fe obtained in sodium chloride electrolytes as a function of chloride concentration 30 minutes after immersion; pH 6.5; aeration*

Figure 2 depicts the current density transients obtained for DHP-Cu at a polarisation potential of $E = E_{corr} + 200$ mV in sodium chloride solutions with different concentrations. All transients show a slight decrease of the currents with increasing polarisation time. However, an increase of the chloride concentration causes an increase of the current densities. Comparable results were found for the current density transients obtained for CuNi10Fe under the same experimental conditions as shown in Figure 3. The transients of CuNi10Fe show higher current densities at the start of the experiment and a stronger decrease than observed for DHP-Cu. At the end of the exposure after 140 h comparable current densities were observed for both materials at the different chloride concentrations.

The corrosion products and the relevant manifestations of corrosion obtained for DHP-Cu and CuNi10Fe after 140 h polarisation at $E = E_{corr} + 200$ mV in sodium chloride electrolytes with different concentrations are summarised in Tables 2 (DHP-Cu) and 3 (CuNi10Fe). The corrosion products were evaluated with regard to their colour and their status of adherence. Pitting attack was rated by determining the obviously deepest pit on the electrode surface, general attack was quantified using the mass loss of the electrodes.

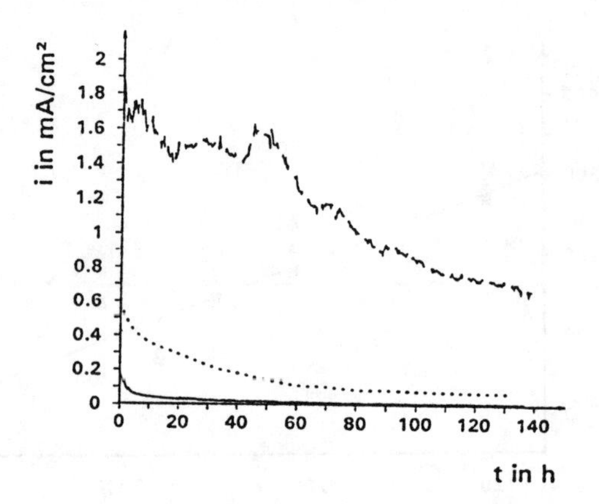

Figure 2 *Current density transients for DHP-Cu measured in NaCl electrolytes at a polarisation potential of $E = E_{Fcorr} + 200$ m V,. pH 6.5; aeration.*
__ 200 mg dm^3 Cl^- ,..... 5.000 mg dm^3 Cl^- , --- 20.000 mg dm^3 Cl^-

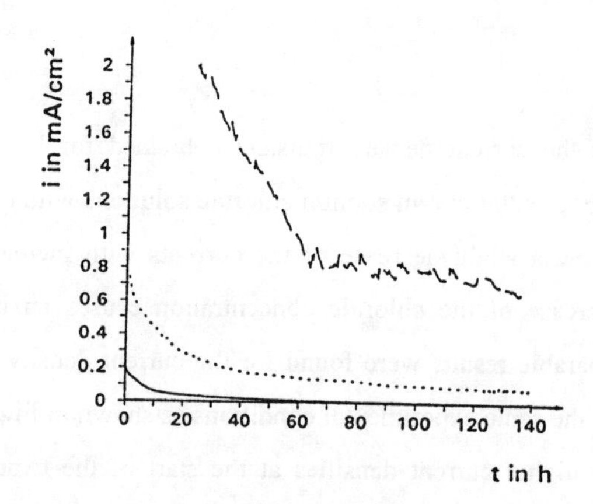

Figure 3 *Current density transients for CuNi10Fe in NaC electrolytesl at a polarisation potential of $E = E_{Fcorr} + 200$ m V,. pH 6.5; aeration.*
__ 200 mg dm^3 Cl^- ,..... 5.000 mg dm^3 Cl^- ,---- 20.000 mg dm^3 Cl^-

No significant influence of the chloride concentration on the appearance of the corrosion products was detectable for DHP-Cu. A loosely adherent dark-red layer was detectable on top of a layer consisting of yellow-green corrosion products independent of the chloride ion concentrations. X-ray diffraction analysis showed that topmost layer consisted of copper(I)-oxide and the bottom layer consisted of copper(I)-chloride[17]. Additionally, in the course of this investigation blue-green corrosion products were obtained at concentrations of $c(Cl^-_{aq}) \geq 1000$ mg dm^{-3} (Table 2).

Table 2 *Evaluation of corrosion products and manifestations of corrosion forDHP-Cu in chloride ion containing electrolytes, (t = 140 h, $E = E_{corr}$ + 200 mV)*

		200 mg/L Cl⁻	500 mg/L Cl⁻	1000 mg/L Cl⁻	2000 mg/L Cl⁻	5000 mg/L Cl⁻	10.000 mg/L Cl⁻	20.000 mg/L Cl⁻
Corrosion Products	dark-red	++	++	++	++	++	++	++
	yellow-green	+	+	+	+	+	+	+
	blue-green	–	–	+	+	+	+	+
	black	–	–	–	–	–	–	–
Manifestations of Corrosion	Maximum pit depths l_{max} in µm	66	110	117	117	75	75	–
	Mass loss Δm_a in g/m²	63	193	194	236	460	1730	3175 3567

– not detectable + adherent layer ++ non-adherent layer

Table 3 *Evaluation of corrosion products and manifestations of corrosion for CuNi10Fe in chloride ion containing electrolytes, (t = 140 h, $E = E_{corr}$ + 200 mV)*

		200 mg/L Cl⁻	500 mg/L Cl⁻	1000 mg/L Cl⁻	2000 mg/L Cl⁻	5000 mg/L Cl⁻	10.000 mg/L Cl⁻	20.000 mg/L Cl⁻
Corrosion Products	dark-red	++	++	++	–	–	–	–
	yellow-green	+	+	+	+	+	+	+
	blue-green	+	+	+	+	+	+	+
	black	–	–	–	+	+	+	+
Manifestations of Corrosion	Maximum pit depths l_{max} in µm	75	115	95	–	–	–	–
	Mass loss Δm_a in g/m²	95	297	216	440	737	1339	697 1769

– not detectable + adherent layer ++ non-adherent layer

Yellow-green and blue-green corrosion products were obtained for CuNi10Fe independent of the chloride concentration. At $c(Cl^-_{aq}) \leq 1000$ mg dm^{-3} these corrosion products were accompanied by a loosely adherent dark-red layer consisting of copper(I)-oxide in an inhomogeneous distribution. This dark-red layer was not detectable any more at higher chloride concentration. Instead of it, black corrosion products were obtained under these conditions (Table 3).

Both general attack and pitting were obtained on DHP-Cu and CuNi10Fe for the various chloride concentrations. For DHP-Cu maximum pit depths $l_{max} \leq 117$ µm were found at $c(Cl^-_{aq}) = 10.000$ mg dm^{-3} indicating a repassivating pitting attack[17]. In parallel, a continuous increase of the mass loss was detectable with increasing chloride concentration (Table 2). The results obtained for CuNi10Fe revealed the same trend (Table 3). However, pitting attack was only detectable at concentrations $c(Cl^-_{aq}) \leq 1000$ mg dm^{-3}. At higher concentrations general attack was the only manifestation of corrosion. Figure 4 depicts the influence of the chloride concentration on the mass loss of DHP-Cu and CuNi10Fe after 140 h at a polarisation potential of $E = E_{corr} + 200$ mV. A comparable mass loss was observable for both materials at concentrations of $c(Cl^-_{aq}) \leq 5000$ mg dm^{-3}. At higher chloride concentrations DHP-Cu showed the more significant corrosion attack.

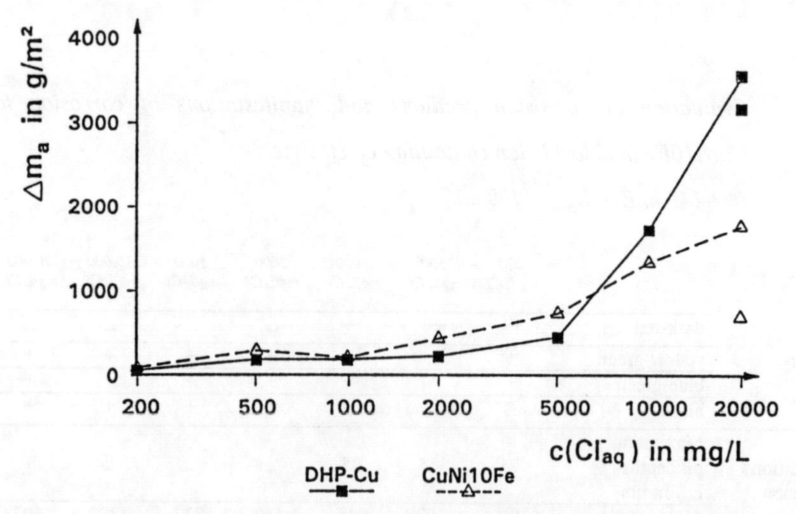

Figure 4 *Mass loss Δm_a of DHP-Cu and CuNiI0Fe after 140 h polarisation at $E = E_{fcorr} + 200$ mV in sodium chloride electrolytes as a function of chloride concentration,, pH 6.5; aeration*

Figure 5 shows the quasi-stationary current density potential curves of DIP-Cu and CuNi10Fe in sodium chloride electrolytes with different chloride concentrations calculated from the current density transients of the potentiostatic series after t = 140 h. All curves show a change in the sign of the current with increasing potential. No significant difference was found for DHP-Cu and CuNi10Fe for the same chloride concentration. Only a moderate slope of the curve was observed for $c(Cl^-_{aq})$ = 200 mg dm^{-3} indicating the formation of a reaction layer that inhibits the anodic partial reaction. A considerably stronger slope was observed at higher chloride concentrations during anodic polarisation. However, a strong variance in the potenial values was found for comparable current densities for $c(Cl^-_{aq})$ = 20.000 mg dm^{-3}

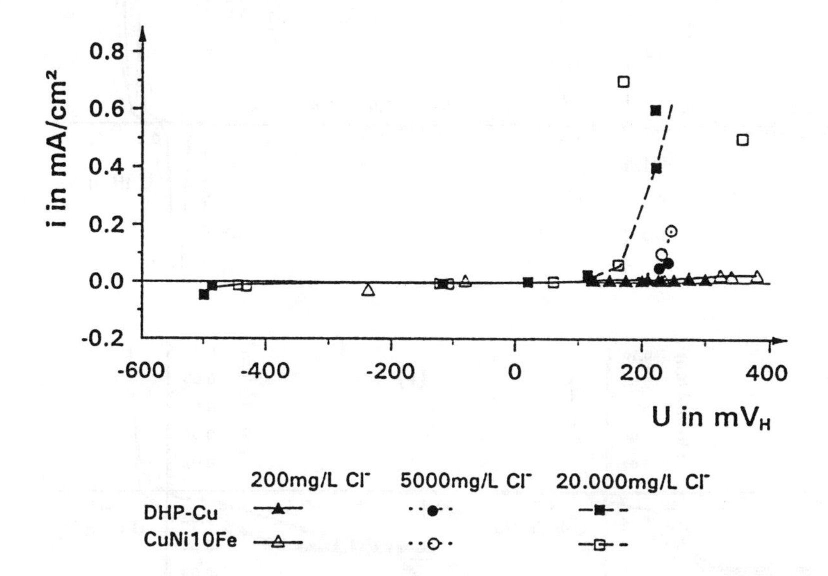

Figure 5 *Quasistationary cuirrent density potential curves of DHP-Cu and CuNi10Fe in sodium chloride for different chloride concentrations calculated from the current transients of the potentiostatic series after t = 140 h*

Figure 6 depicts the current density potential curves of DIP-Cu and CuNi10Fe in electrolytes with $c(Cl^-_{aq})$ = 200 mg dm^{-3}, 5000 mg dm^{-3} and 20.000 mg dm^{-3}.

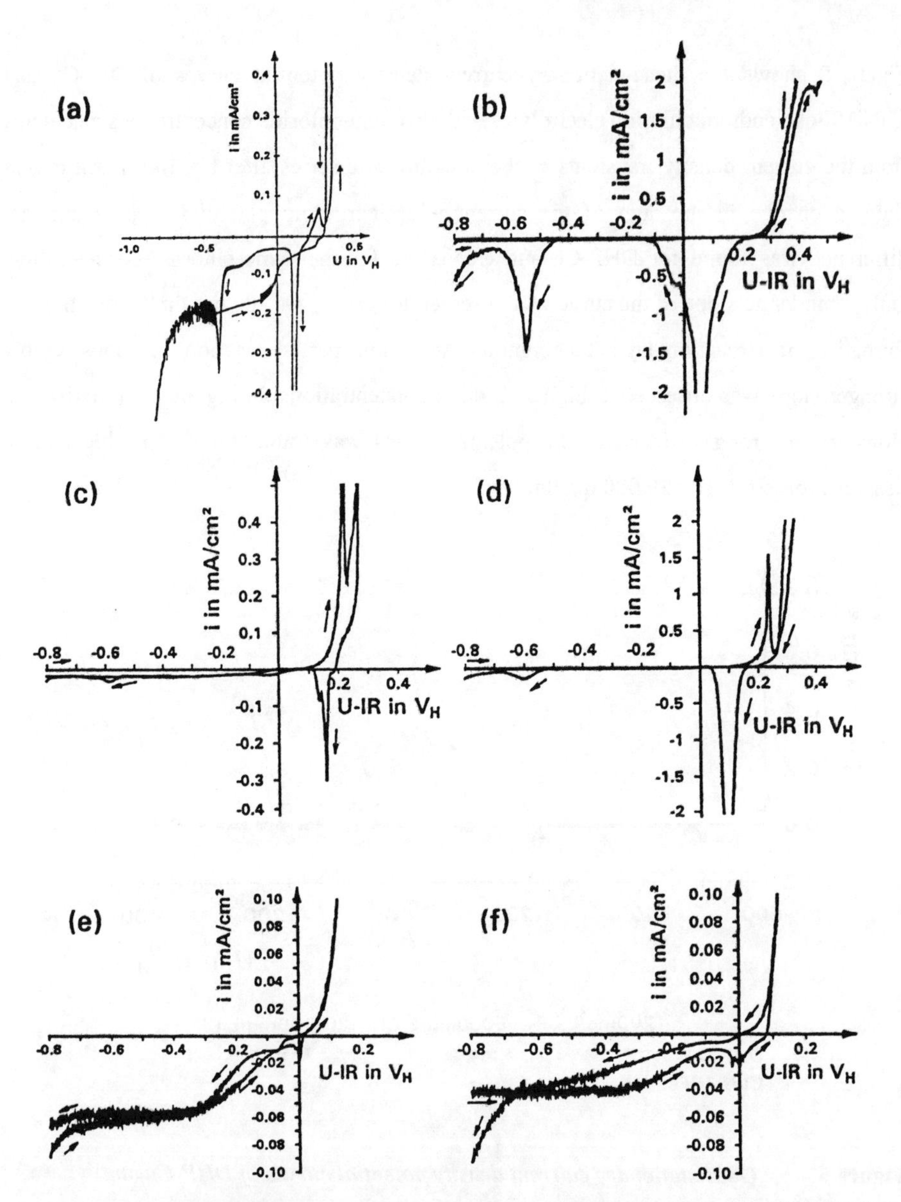

Figure 6 *Potentiodynamic current density potential curves of DHP-Cu and CuNi10Fe in sodium chloride with different chloride concentrations; pH 6.5; aeration; scan rate: 0.01 mVs^{-1}; start potential: -800 mV_H*

(a) DHP-Cu; $c(Cl^-_{aq})$ = 200 mg/L; U_{th} = 210 mV_H

(b) CuNi10Fe; $c(Cl^-_{aq})$ = 200 mg/L; U_{th} = 185 mV_H

(c) DHP-Cu; $c(Cl^-_{aq})$ = 5000 mg/L; U_{th} = 64 mV_H

(d) CuNi10Fe; $c(Cl^-_{aq})$ = 5000 mg/L; U_{th} = 99 mV_H

(e) DHP-Cu; $c(Cl^-_{aq})$ = 20.000 mg/L; U_{th} = 10 mV_H

(f) CuNi10Fe; $c(Cl^-_{aq})$ = 20.000 mg/L; U_{th} = 10 mV_H

The curves start in the range of the hydrogen evolution, then a first plateau is observed that can be attributed to the oxygen reduction reaction with water[21-23]. At more positive potentials the cathodic current shows a further decrease. In this potential range thin tarnishing layers were detected on the material surfaces after performing potentiostatic experiments[17]. This anodic partial reaction is superimposed by an oxygen reduction reaction under proton consumption[21]. The threshold potential E_{th} decreases with increasing chloride concentration for both materials and it is situated in the range of the free corrosion potentials in the different electrolytes (see Figure 1).

At potentials positive of the threshold potential a strong increase of the current was observed due to the oxidation of copper to copper(I)-ions[24,25]. At more positive potentials copper(II)-corrosion products would be obtained. An oxidation peak and a corresponding reduction peak were observed in Figures 6a, c and d in a potential range of 50 mV_{NHE} to 300 mV_{NHE}. These peaks are attributed to the oxidation and reduction of copper(I)-chloride[26]. In Figure 6b only the reduction peak was found. These peaks were not detectable any further at chloride concentrations of $c(Cl^-_{aq})$ = 20.000 mg dm^{-3} (Figure 6e and f). A second reduction peak was detectable in the potential range of -400 mV_{NHE}. to -600 mV_{NHE} in the reverse scan for both materials of $c(Cl^-_{aq})$ = 200 mg dm^{-3} and 5000 mg dm^{-3} (Figures 6a-d). This peak corresponds to the reduction of copper(I)-oxide according to the literature[24,25,27]. This corrosion product was also identified by X-ray diffraction analvsis after performing potentiostatic experiments. In Figures 6e and f, this peak did not occur.

The corrosion products obtained on CuNi10Fe after performing the potentiostatic series at $E = E_{Fcorr}$ + 200 mV in electrolytes with different chloride concentrations were investigated by Scanning Electron Mcrosocpy (SEM) and Wave Length Dispersive Analysis (WDX). Figure 7 shows the scanning electron micrographs obtained from cross sections of the CuNi10Fe electrodes, Figure 8 gives selected point analyses. The non-attacked material is situated in the top right corner of the micrograph, the epoxy resin in the bottom left corner, respectively. A layer consisting of compact corrosion products of about 100 μm thickness was obtained for $c(Cl^-_{aq})$ = 1000 mg dm^{-3} (Figure 7A). A material analysis of the corrosion products was performed close to the material surface.

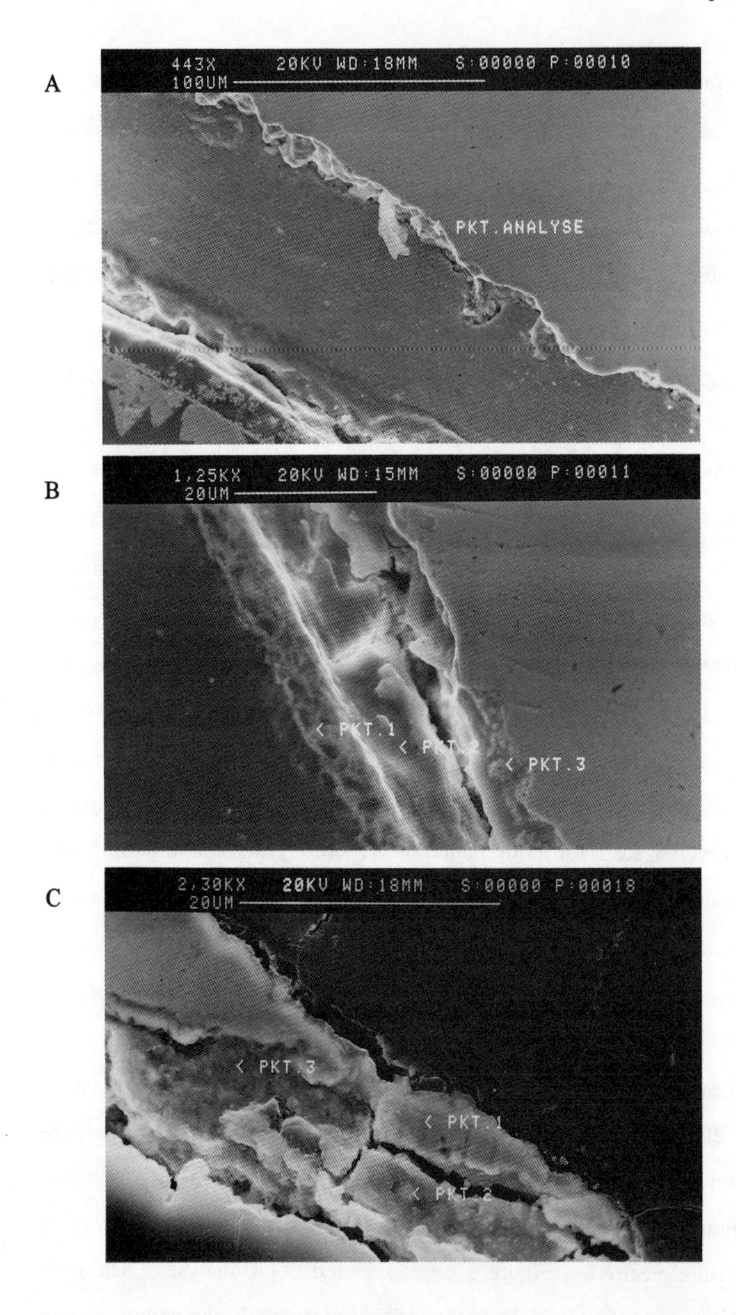

Figure 7 *Scanning electron micrographs obtained from cross sections of CuNi10Fe electrodes after performing the potentiostatic series for 140 h at* $U = U_{FKorr} +$ *200 mV in sodium chloride electrolyte with different chloride concentrations*
A: $c(Cl^-_{aq}) = 1000$ *mg/L* *B:* $c(Cl^-_{aq}) = 5000$ *mg/L*
C: $c(Cl^-_{aq}) = 20.000$ *mg/L*

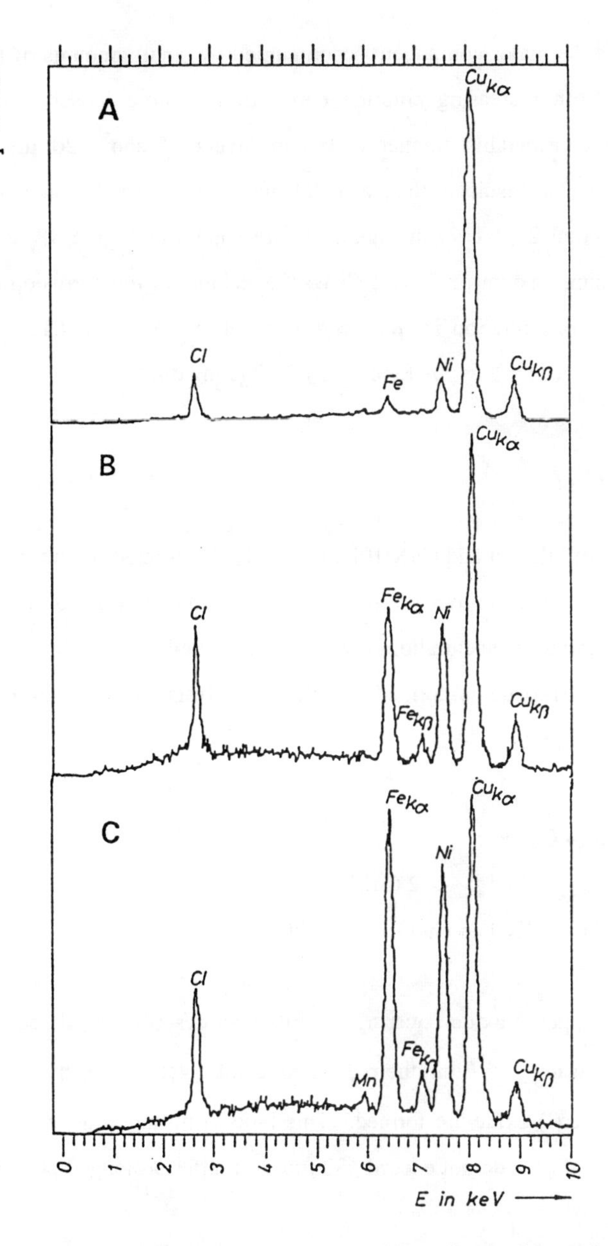

Figure 8 *Wavelength dispersive point analyses obtained from the micrographs shown in Fig. 7*
A: Point analysis of Fig. 7A; $c(Cl^-_{aq})$ = 1000 mg/L
B: Analysis of point 1 in Fig. 7B; $c(Cl^-_{aq})$ = 5000 mg/L
C: Analysis of point 2 in Fig. 7C; $c(Cl^-_{aq})$ = 20.000 mg/L

Copper was identified as the major component accompanied by small amounts of Fe, Ni and Cl (Figure 8A). With increasing chloride concentration more loosely adherent corrosion products and considerably thinner corrosion layers of about 20 μm were obtained (Figures 7B, 7C). The results of the elemental analysis performed at point 1 (Pkt. 1) in Figure 7B and at point 2 (Pkt. 2) in Figure 7C are shown in Figure 8B and 8C, respectively. An increasing amount of Fe and Ni was detectable in the corrosion layers with increasing chloride concentration in the electrolyte. The amount of Cl was also considerably higher as observed in Fig. 8A for $c(Cl^-_{aq}) = 1000$ mg dm^{-3}.

4. CONCLUSIONS

The corrosion behaviour of DIP-Cu and CuNi10Fe is mainly determined bv the influence of chloride ions in various concentrations on the formation of copper corrosion products. In electrolytes with chloride concentrations typical for potable water an adherent protective film of copper(I)-oxide is formed via the hydrolysis of the bottom layer consisting of copper(I)-chloride[17].

$$Cu \rightarrow Cu^+ + e^- \quad (1)$$

$$2Cu^+_{aq} + 2\,Cl^-_{aq} \leftrightarrow 2\,CuCl \quad (2)$$

$$2CuCl + H_2O \leftrightarrow Cu_2O + 2H^+ + 2Cl^- \quad (3)$$

The film consisting of copper(I)-oxide covering the pits was described by Lucey to be typical for pitting in cold water[28]. Within the pits an adherent layer of copper(I)-chloride and large crystals of copper(I)-oxide are formed. This type of pitting is a repassivating pitting. The maximum pit depths do not exceed 130 μm even after prolonged polarisation times.

With increasing chloride concentration the completing properties of copper(I)-chloride complex ions become dominant. The formation of a protective layer of copper(I)-oxide via hydrolysis of copper(I)-chloride is prevented at high chloride concentrations e.g. according to the following reaction scheme describing the formation of $CuCl_2^-$ complex ions:

$$Cu \rightarrow Cu^{+} + e^{-} \quad (1)$$

$$Cu^{+} + 2Cl^{-} \rightarrow CuCl_2 \quad (4)$$

The equilibrium constant of reaction (4) can be described by the following equation:

$$K = [CuCl_2^{-}]/[Cu^{+}][Cu^{-}]^2 \quad (5)$$

With K' = K . [Cu⁺l the following relationship is obtained:

$$K' [Cl^{-}]^2 = [CuCl_2^{-}] \quad (6)$$

Equation (6) indicates a quadratic influence ofthe chloride concentration on the formation ofthe complex ion $CuCl_2^{-}$.

When the corrosion process of a copper alloy is mainly determined by reaction (4), the material surface keeps active. On DHP-Cu no adherent reaction layers are formed, whilst nickel and iron support the formation of adherent reaction layers on CuNi10Fe at high chloride concentrations.

References

1. B. B. Moreton, CDA (Copper Development Association) Seminar 'Copper Alloys in Marine Environments', Birmingham, UK, April 1985, Paper No. 1.
2. W. W. Kirk, T. S. Lee, R. 0. Lewis, CDA (Copper Development Association) Seminar 'Copper Alloys in Marine Environments', Birmingham, UK, April 1985, Paper No. 16.
3. R. F. North and M. J. Pryor, *Corr. Sci.*, 1970, **10**, 297.
4. J. M. Popplewell, R. J. Hart and J. A. Ford, *Corr. Sci.*, 1973, **13** , 295.
5. C. Kato, B. G. Ateya, J. E. Castle and H. W. Pickering, *J. Electrochem. Soc.*, 1980, **127**, 1890.
6. C. Kato, J. E. Castle, B. G. Ateya and H. W. Pickering, *J. Electrochem. Soc.*, 1980, **127**, 1897.
7. C. Kato and H.W. Pickering, *J. Electrochem. Soc.*, 1984, **131**, 1219.

8. D. D. MacDonald, B. C. Syrett and S. S. Wing, *Corrosion*, 1978, **34**, No. 9.
9. J. M. Krougman and F. P. ljsseling, Proceedings of the 4th International Congress on Marine Corrosion and Fouling, Antibes, France, June 1976.
10. F. J. Kievits and F. P. ljsseling, *Werkst. Korr.*, 1972, 23.
11. R. G. Blundy and M. J. Pryor, *Corr. Sci.*, 1972, **12**, No. 1.
12. DECHEMA-Werkstofftabelle, 'Meerwasser', Verlag Chemie, Weinheim, September 1975.
13. F. M. Al-Kharafi and Y. A. El-Tantavy, *Corr. Sci.*, 1982, **22**(1).
14. R. May, *J. Inst. Metals*, 1954, **32**, 65.
15. M. Pourbaix, *J. Electrochem. Soc.*,1976, **123**, 27.
16. W. R. Fischer and B. Füssinger, Proceedings of 12th Scandinavian Corrosion Congress & EuroCorr 92, Dipoli, Finland, 1992, 769 - 778.
17. H. Siedlarek and B. Füssinger, I. Hänssel and W. R. Fischer, *Werkst. Korr.*, 1994, **45**, 654.
18. W. Fischer and W. Schwenk, 'Electrochemical Corrosion Testing', Eds. E. Heitz, J. C. Rowlands, F. Mansfeld, DECHEMA Monographien, 1985, **101**, 53 8.
19. W. R. Fischer, *Werkst. Korr.*, 1988, **39**, 538.
20. DIN 50 905, Corrosion of metals, corrosion testing, general guidance, Beuth Verlag Berlin, September 1985.
21. W. Fischer and W. Siedlarek, *Werkst. Korr.*, 1979, **30**, 695.
22. C. Müller and L. N. Nekrassov, *Electrochim. Acta*, 1964, **9**, 1015.
23. C. Müller and L. N. Nekrassov, *Electrochim. Acta*, 1965, **9**, 282.
24. C. D. Burke and T. G. Ryan, *J. Electrochem. Soc.*, 1990, **137**, 1358.
25. H. H. Strehblow and H.-D. Speckmann, *Werkst. Korr.*, 1984, **35**, 512.
26. W. R. Fischer, in preparation.
27. J. G. N.Thomas and A. K. Tiller, *Br. Corr. J.*, 1972, **7**, 256.
28. V. F. Lucey, *Werkst. Korr.*, 1975, **26**, 185.

8. MICROBIAL CORROSION: EFFECT OF MICROBIAL CATALASE ON THE OXYGEN REDUCTION

J.P. Busalmen, M.A. Frontini and S.R. de Sánchez.

Corrosion Division, Instituto de Investigaciones y Tecnología de Materiales (INTEMA), Fac. de Ingeniería, U.N. de Mar del Plata, Juan B. Justo 4302, (7600) Mar del Plata, Argentina

1. INTRODUCTION

A very common deterioration form for metallic materials is the one resulting from the growth of bacteria attached to the metal surface. The initial attachment of an individual bacterium leads to the formation of a biofilm matrix of variable composition produced by them. The biofilm grows incorporating other organic and inorganic materials and other organisms from the environment[1].

Amongst the most widely studied micro-organisms associated with biocorrosion are the anaerobic sulphate reducing bacteria, for which the mechanism of action has been described earlier[2]. Due to the presence of hydrogenase enzyme, the capacity of these bacteria to remove cathodically produced hydrogen from the metal was thought to be one of the reasons for their corrosivity [3,4]. However, there is very little information about the mechanism by which aerobic bacteria affect the corrosion behaviour, and this is the aim of this study.

The present work is part of a failure analysis of a heat exchanger in a power station. The inner surface of the tubes was found to be covered with a slime film. The development of biofilms on electric power plants heat exchangers tubes, usually made of copper alloys, causes two main problems: a severe reduction of the heat transfer capability and the enhancement of the corrosion rate[5-8]. Several authors[9-12] noted that when a biofilm is present on the metal surface, the oxygen reduction is affected. Sánchez *et al.*[8]

demonstrated that the slime layer acts as a partial barrier to the diffusion of oxygen and that the kinetics of oxygen reduction is changed from activation control to diffusion control through the slime layer and that the electron transfer reaction at the slime-metal interface is greatly enhanced. Mollica *et al.*[9,10] and Scotto *et al.*[11,12] showed similar results for stainless steel, nickel, titanium and cooper-nickel alloys. Those results were attributed to an enzymatic catalysis affecting the oxygen reduction kinetics. The electrochemical reduction of oxygen on metals is of great practical interest since it is the cathodic process in the corrosion of metals in aerated aqueous environments. For copper alloys in sea water, this is also the corrosion rate controlling reaction. On copper the reaction proceeds by a four electron mechanism and it is apparent that changes in surface oxidation and electrolyte composition have a great influence on the kinetics of oxygen reduction [13-17]. The exchange of four electrons during the O_2 reduction may arise either from direct reduction without the formation of intermediate peroxide, or by a sequential mechanism with elimination of adsorbed peroxide:

$$O_{2\,(sol)} \xrightarrow{k_I} O_{2\,(ads)} \xrightarrow{k_{II}} H_2O_{2\,(ads)} \begin{cases} \xrightarrow{k_{III}} OH^- \\ \underset{k_V}{\overset{k_{VI}}{\rightleftarrows}} H_2O_{2\,(sol)} \end{cases}$$

On copper, Vazquez *et al.*[16] found that the observed yield of peroxide is small and proposed that $k_{III} >> k_V$, so that, the peroxide reduction on Cu is very fast, and a single four electron reduction wave is observed. The same authors [17], also demonstrated that in the case of the hydrogen peroxide reduction, the presence of Cu_2O is necessary. This oxide is soluble in chloride solutions and in this case the hydrogen peroxide reduction occurs at much negative potentials.

2. EXPERIMENTAL METHODS

The electrodes for the electrochemical measurements were made of tubing pieces of Aluminium Brass (ASTM B111) mounted with fast curing epoxy resin on appropiated Teflon holders[1]. Before each experiment the electrode surface was smoothed with 600 grit emery paper and then mirror polished with 0.3 μm alumina powder.

The strains used in the present work were isolated from the gelatinous deposit present on the inside walls of the condenser tubes. These samples were collected using sterile swabs and were them developed in a culture medium appropiated for marine bacteria. The composition of the culture medium was: Lab Lemco (Merck) 0.1% (P/V), yeast extract (Sigma) 0.2% (P/V), and peptone (Sigma) 0.5% (P/V), disolved in artificial sea water.

For the isolation and purification of the strains, standard bacteriological techniques were employed [18,19].

The original sample was plated out to obtain pure cultures, which were passed through a series of identification tests. These were: Gram staining, oxidation-fermentation (Hugh and Leifson), motility, catalase, gelatine liquefaction, oxidase, pigmentation and morphology.

2.1 Weight loss experiments

The selection of the strain was based in weight loss experiments. Rectangular samples (1 x 2 cm) were cut from flattened aluminium brass condenser tubes and abraded with 600 grit emery paper. Then they were carefully washed with distilled water, dried and weighted. These samples were exposed during 10 days to the culture medium inoculated with the different isolated strains. After this period of time, the samples were carefully washed, dried and weighted again.

2.2 Electrochemical Experiments

In order to test the effect of bacteria, electrochemical experiments were performed. The potentiostat used was a LYP M-5, La Plata, Argentina. The growth of bacteria was carried out at 32 °C during 18-22 hs. in the media described before.

To study the effect on the corrosion behaviour of the culture medium, blank experiments were performed with sterilised media.

2.2.1 Polarisation Curves Polarisation curves were performed in two conditions:

a) with a culture at the begining of the stationary phase.
b) with a bacteria-free culture (BFC) obtained by filtration through a 0.22 μm pore diameter membrane filter, while keeping the system temperature below 4 °C.

The absence of bacteria in BFC was verified by plating several portions in solid culture media and checking that no colonies grow. To control the effect of temperature on the BFC, a portion of it was maintained to 80 °C during 15 min. and used afterwards.

The test cell used in these experiments was a three electrode electrochemical cell. The potential was measured against a Ag/AgCl-KCl saturated reference electrode. Before conducting a polarisation curve, the electrolyte was saturated with air, to ensure a common base of comparison for the cathodic process.

The culture and its filtrate (BFC) were tested for catalase activity by adition of 35% H_2O_2 solution [20].

2.2.2 Electrochemical production of H_2O_2 The electrochemical production of hydrogen peroxide was tested in an specially adapted spectrophotometer quartz cell. An aluminium brass electrode was arranged in a lateral position to prevent interrupting the light beam; opposite to it, a Pt wire acted as a counter electrode; a salt bridge to provide the connection with the reference electrode camera (Ag/AgCl/KCl sat) was used. The electrode was polarised to -0.30 V in a 3.5% NaCl solution and absorption spectra of the products were registered between 200-800 nm during 30 min. A solution of 1.6 mM H_2O_2 in NaCl 3.5% was used as the standard solution.

Enzymatic degradation of polarisation products was assayed by adding catalase to the final solution of previous experiment. An absorbance vs. time spectrum was recorded at 205 nm during 4 minutes.

The catalase solution (0.5 mg cm^{-3} - NaCl 3.5%) was prepared with catalase (Sigma) from bovine liver. The spectrophotometer was a Shimadsu UV-160A UV-Visible.

2.2.3 Reduction of enzymatically produced oxygen An aluminium brass electrode was placed in on air-proof electrochemical cell with a capacity of 2.5 cm^3, with an Ag/AgCl-KCl saturated reference electrode and a platinum counter electrode. The cell was filled with deareated 40 mM H_2O_2-NaCl 3.5% solution and the electrode was polarised at -0.30 V; reduction currents were measured. When a constant current was reached, 50 µl of catalase solution (0.5 mg cm^{-3} in NaCl 3.5%) were added and the current continued being registered for 15 more minutes. Solutions were deaerated with argon gas (4 N quality).

3. RESULTS

Weight loss experiments were performed with the characterised strains isolated from the condenser, in order to select the one with the biggest effect on the corrosion rate for the material tested. Table 1 shows the results obtained with two *Pseudomona* strains and *Micrococcus sp.* Some other strains were tested with no significative weight loss. The results show that strains of the *Pseudomona* genera have the strongest corrosive action, so, the *Pseudomona sp.* 1 was selected for this study.

Table 1 *Weight loss values for Al-Brass in the presence of different strains of bacteria after 10 days of exposure (mg cm^{-2}).*

Media	Sterile	Inoculated with *Pseudomona sp* (r)	Inoculated with *Pseudomona sp* (s)	Inoculated with *Micrococcus sp*
Weight loss	1,2	16,8	15,5	1,5

The influence of both, bacterial culture and bacteria free culture (BFS), on the cathodic polarisation behaviour of the aluminium brass is shown in the Figure 1.

The corresponding polarisation curves are shown compared to the blank, recorded in sterilised media. In both cases, the oxygen reduction rate is enhanced, but the increase in current is much higher for BFC.

The enhancement of O_2 reduction current produced by the BFC disappears after heating to 80 °C and the current drops to the values of the blank as can be seen in Figure 2. Spectrophotometrical tests were performed to detect the formation of hydrogen peroxide as intermediate product.

The result of the cathodic polarisation of the Al-Brass electrode to a -0.30 V potential in a chloride containing solution is presented in Figure 3. Successive absorption spectra of polarisation products recorded over 30 minutes show that the absortion increased in between the range of 200-240 nm with a maximun at 205 nm which corresponds to hydrogen peroxide, taking a standard solution as reference.

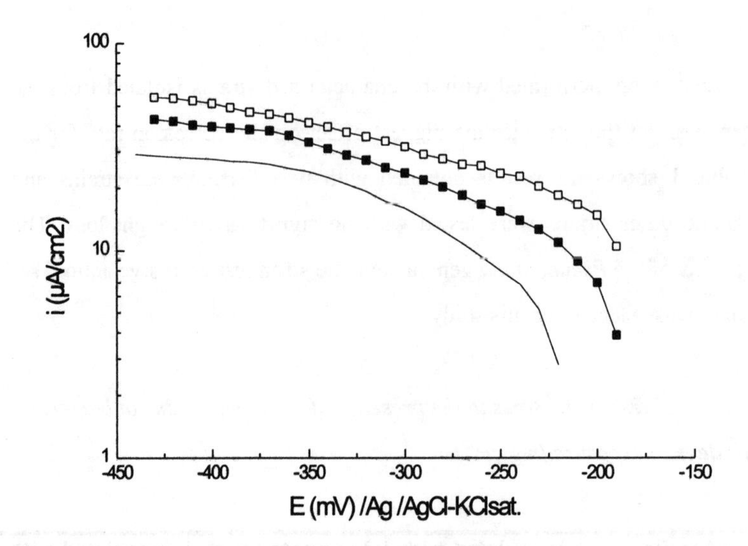

Figure 1 *Polarisation curves of Al-Brass in (■)stationary-phase culture and (□) BFC of Pseudomona sp. 1. (—) Blank in sterilised medium.*

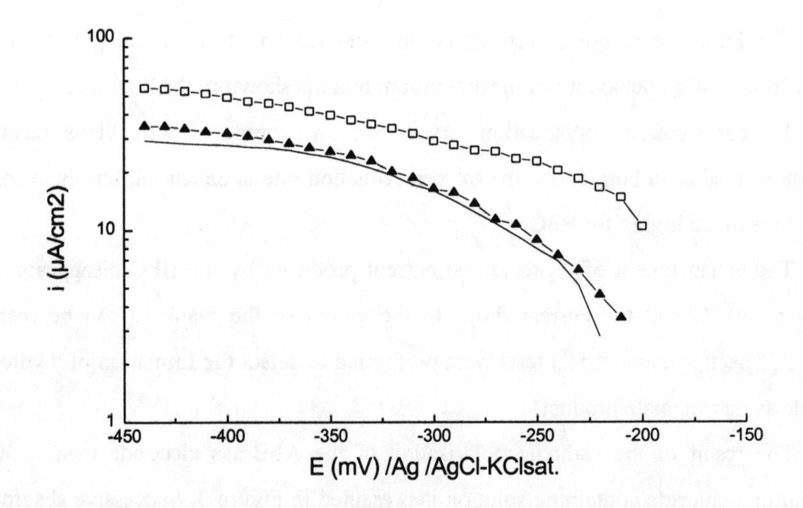

Figure 2 *Polarisation curves of Al-Brass in aerated BFC of Pseudomona sp. 1 (□) before and (▲). Polarisation after heating at 80 °C for 15 minutes. (—) Blank in sterilised medium*

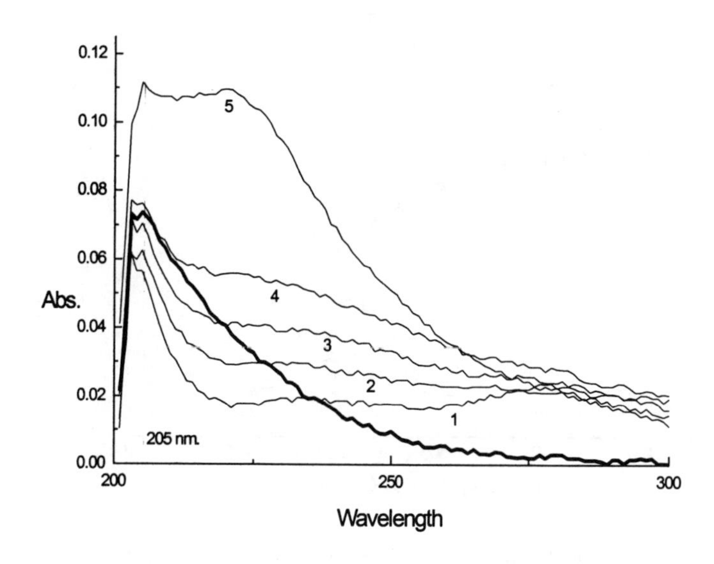

Figure 3 *Absorption spectra of polarisation products of Al-Brass polarisation in NaCl 3.5% at -0.30 V after: (1) 1 minute, (2) 5 minutes, (3) 10 minutes, (4) 15 minutes, and (5) 30 minutes. (—) Absorption spectra of standard solution (1.6 mM H_2O_2 - NaCl 3.5%)*

The H_2O_2 reduction rate is very low at the established potential compared to the O_2 reduction rate as was probed by polarisation curves in 1 mM H_2O_2-NaCl 3.5% deaerated with Ar (Figure 4). For this experiment the O_2 reduction curve was measured in a solution saturated with pure oxygen (~ 1.3 mM).

In order to verify if catalase could be active in the prsence of the hydrogen peroxide formed in the above experiment, its activity was spectrophotometrically followed after adding the enzyme. Figure 5 shows the fast decrease of absortion at 205 nm due to consumption of hydrogen peroxide by catalase. The action of catalase is first-order in both substrate and enzyme concentration, and at the low peroxide concentration a smooth consumption rate can be observed after 25 seconds of enzyme addition.

Assuming that the presence of catalase in stationary-phase culture and in BFC can modify the oxygen reduction currents, an experiment was designed to elucidate whether

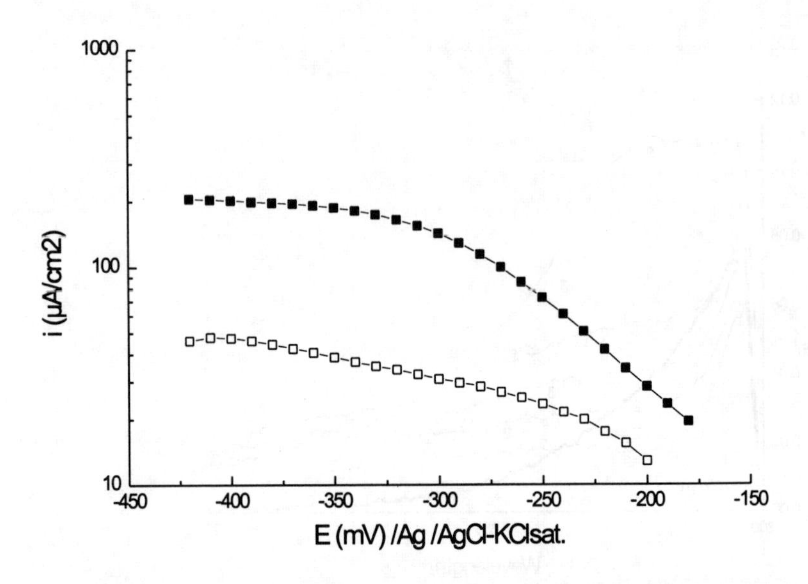

Figure 4 *Polarisation curves of Al-Brass in (□) deaerated 1 mM H_2O_2 - NaCl 3.5% and, (■) NaCl 3.5% solution saturated with pure oxygen*

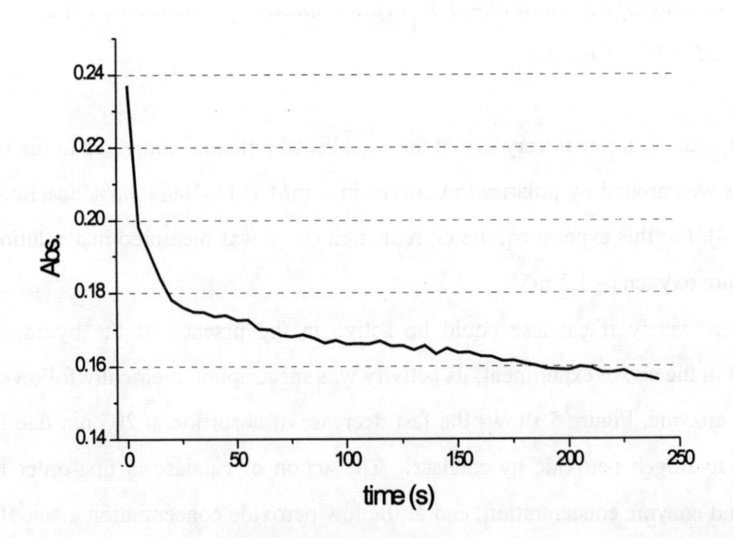

Figure 5 *Enzymatic consumption of hydrogen peroxide after addition of catalase solution to final polarisation products of Al-Brass (Figure 3).*

the effect of the catalase enzyme is an enhancement of the peroxide reduction rate at the working potential or, otherwise, the production of additional oxygen with the subsequent increment in its reduction current.

The results in Figure 6 show that a steady state reduction current can be reached by polarisation in a peroxide concentrate (40 mM) solution (zone I); if at that point a catalase solution is added, the current decreases with time due to peroxide enzymatic decomposition (zone II); after a few minutes current raises again because the reduction of the liberated oxygen (zone III). Oxygen bubbles formation made necessary high peroxide initial concentration to reach a final dissolved oxygen quantity which could be detected electrochemically.

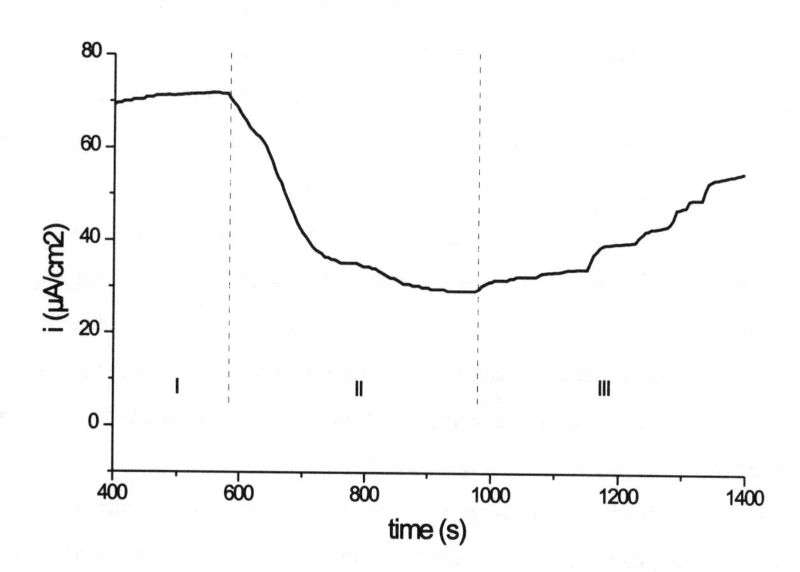

Figure 6 *Electrochemical reduction of enzymatically produced oxygen after addition of 50 µl of catalase solution to the deaerated 40 mM H_2O_2 - NaCl 3.5% initial solution. (I) Hydrogen peroxide steady state current phase, (II) Peroxide reduction current decrease phase due to catalase activity, and (III) Oxygen reduction current phase*

4. DISCUSSION

The influence of bacteria on the cathodic reaction has been intensively studied for the sulphate reducing bacteria[2-4]. However, the modification introduced by aerobic bacteria has received much less attention.

Oxygen reduction may be monitored using electrochemical measurements; Mollica *et al.*[9,10] reported that the oxygen reduction current increased drastically during time exposure to natural sea water over a few days, this increment was related to aerobic bacterial growth and biofilm formation. Adding an enzymatic inhibitor, Scotto. *et al* [11,12] showed that the effect of living cells on the corrosion process is probably due to enzymatic catalysis affecting the oxygen reduction kinetics, however nothing can be concluded from these results about the implicated mechanism.

The results in Figure 1 showed the increase in the reduction current in the presence of bacteria according with the previous reports mentioned above. However the working conditions were different. In this work, the time of metal exposure to the stationary-phase culture when the effect was detected, was too short to allow the biofilm formation. This observation led us to conclude that biofilm formation is not necessary to promote current increments. In fact, using a BFC obtained from a stationary-phase culture, higher reduction currents can be obtained (Figure 1). This result suggests the presence of some bacterial product able to modify the kinetics of the O_2 reduction reaction independently of the presence of bacteria. This bacterial product seems to be heat-sensitive since the catalytic effect disappears after a 15 min. incubation at 80 °C; the cathodic current returns to the blank values (Figure 2).

The catalysis of the cathodic process by enzymatic metabolites has been previously proposed for sulphate reducing bacteria[3,4], but very little information concerning biological catalysis in aerobic systems is available.

The *Pseudomona sp.* exert a positive effect on the catalase reaction and the presence of this enzyme could be tested in stationary-phase culture and BFC. Hydrogen peroxide is the catalase substratum and its formation during cathodic polarisation was detected by spectrophotometry; the results of which are summarised in Figure 3. This experiment suggests that on aluminium brass oxygen reduction occurs *via* less than four electrons in chloride containing media, and that the hydrogen peroxide formed easily desorbs into the solution. At the selected potential, the electrochemical reduction rate for H_2O_2 on Al-Brass is very low as can be appreciated in Figure 4, resembling its behaviour on copper in similar media (Vázquez *et al.* [17]). The O_2 reduction, otherwise, is not impeded in these conditions.

Taking into account reactions in the scheme below, the H_2O_2 could be decomposed by catalase action (Figure 5), thus, additional oxygen would be produced near the surface and higher oxygen reduction currents should be measured.

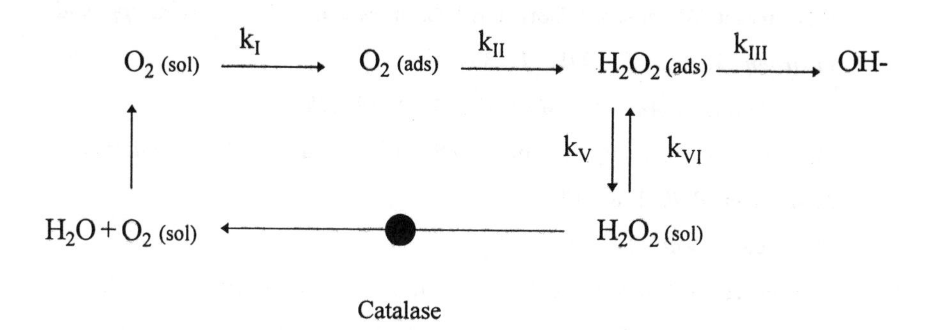

The results shown in Figure 6 correspond to an experiment that simulates the action of catalase in a closed system where the access of oxygen is limited and the oxygen formed remains in the medium. This result allows us to propose that the O_2 reduction rate enhancement in the presence of bacterial culture or BFC, is due to the reduction of the additional oxygen liberated by enzymatic work. These results agree with previous reports describing an auto-catalytic oxygen reduction mechanism in the presence of an aerobic bacteria biofilm[10].

Schellhorn *et.al.*[20] reported induction of hydroperoxidase in *E. Coli* by oxidative stress, it is likely that the catalase production may be incremented by peroxide accumulation under the biofilm. It is not possible, at present, to quantify the relative importance of catalase contribution in the total observed effect. However our results show that this enzyme participates actively in the oxygen reduction cycle in systems involving *Pseudomona sp.* in contact with copper alloys.

Additional information could be obtained by isolation of bacteria enzyme from culture medium and further characterisation of its working conditions. This work is currently in progress in our laboratory.

References

1. A.H.L Chamberlain, *Bioflims, Science and Technology*, L.F. Melo et.al.(Eds.). Kuwet Academic Publishers., 207 (1992).
2. C.A.H. von Wolzogen Kuhr and L.S.Van der Vlugt, 1934, **18**, 147.
3. R.D. Bryant, W. Jansen, J. Boivin, E.J. Laishley and J.W. Costerton, *App. and Environm. Microbiol.*, 1991, **57**, 2804.
4. W.A. Hamilton, *Ann. Rev. Microbiol.*, 1985, **39**, 195.
5. Sh.G Berk, R. Mitchell, R.J. Bobbie, J.S. Nickels and D.C. White, *Internat. Biodeterior. Bull.*, 1981, **17**(2).
6. R.O. Lewis, *Mat. Perf.*, 1982, 31.
7. C. Manfredi, S. Simison and S.R. de Sánchez, *Corrosion*, 1987, **43**, 458.
8. S.R.de Sánchez and D.J. Schiffrin, *Corrosion*, 1985, **41**, 1.
9. A. Mollica, A. Trevis, E. Traverso, G. Ventura, V. Scotto, G. Alabiso, G. Marcenaro, V. Montini, G. De Carolis and R. Dellepiane, Proceedings of the 6th International Congress on Marine Corrosion and Fouling, Athens, Greece, 1984.
10. A. Mollica, G. Ventura, E. Traverso and V. Scotto, *Int. Biodet.*, 1988, **24**, 221.
11. V. Scotto, R. Di Cintio and G. Marcenaro, *Corr. Sci.*, 1985, **25**, 185.
12. V. Scotto, G. Alabiso and G. Marcenaro, *Bioelectroch. and Bioenerg.*, 1986, **16**, 347.
13. G. Faita, G. Fiori and D. Salvatore, *Corr. Sci.*,1975, **15**, 383.
14. H.P. Lee and K. Nobe, *J. Electrochem. Soc.*, 1986, **133**, 2035.
15. K. Balakrishnan and V.K. Venkalesan, *Electrochem. Acta.*, 1979, **24**, 131.
16. M.V. Vázquez, S.R.de Sánchez, E. Calvo and D.J. Schiffrin, *J. Electroanal.Chem.*, 1994, **374**, 189.
17. M.V.Vázquez, S.R.de Sánchez, E. Calvo and D.J. Schiffrin, *J. Electroanal. Chem.*, 1994, **374**, 179.

18. C.H. Collins, Microbiolgical Methods. Butterworths,London (1967).

19. Bergey, Manual of Determinative Bacteriology.8th Edition., The Williams and Wilkins Company, Baltimore (1974).

20. H.E. Schellhorn, *FEMS Microbiol. Letters.*, 1995, **131**, 113.

9. NEW ANTICORROSIVE PIGMENTS*

Th.Skoulikidis, P.Vassiliou, S.Vlachos

Department of Materials Science and Engineering, Faculty of Chemical Engineering, National Technical University of Athens, 9, Iroon Polytechniou Street, Athens 157 80, GREECE

Abstract

There are some methods of protection of metals against corrosion that act by decreasing the corrosion potential and current. For the same purpose we introduced a new system using n-semiconductors as pigments in polymer vehicles. The n-semiconductors, having the predisposition to offer electrons, impose a type of cathodic protection, they do not get exhausted as materials and as electron donors, they act as anti-UV sensitisers and increase the anticorrosive properties of known paints. They also replace part of the needed extemal cathodic potential if cathodic protection and anticorrosive paints are simultaneously used.

1. INTRODUCTION

Among the several methods of protection of metals against corrosion there are some, such as cathodic protection by sacrificial anodes (in indirect or direct contact with the metal surface), by extemal imposed current, by sacrificial metal powders in anticorrosive paints and by using atmospheric electricity [needle-diodes method[1-3]], that act by decreasing the corrosion potential and current. Promoting reduction reactions, they slow down the oxidation rate of metals. For the same purpose we introduced a new system using n-semiconductor pigments in polymeric vehicles. The n-semiconductors, having the predisposition to offer electrons, could impose a type of cathodic protection; they do not get exhausted and, depending on their structure, they can protect their polymeric vehicles from UV light degradation acting as sensitisers.

*Patent # 1000449 Int. Cl. CO9D 5/08

2. EXPERIMENTAL DETAILS

2.1 Materials, Specimens and Experimental Procedures

The following n-semiconductors, ZnO, $A1_2O_3$, Fe_2O_3, Fe_3O_4 were prepared by special preparation methods in order to give them pronounced n-semiconductor properties[4,5 6] (Patent No 1000449, Int.Cl.C09D 5/08). It must be emphasised that in order for a substance to be an n-semiconductor and especially to have high semiconductiing intensity depends not only on its nature but principally on the lack of stoichiometry, depending on its method of preparation . Thus $A1_2O_3$, following its method of preparation, can be an insulator or an n- or p-semiconductor.

As polymeric vehicles epoxy resin (E.R), Coal Tar Epoxy (C.T.E.), Epoxy with Zinc powder (Z.R.E.), Chlorinated Rubber (C.R.) and Red Mud, (R.M) [7,8] (an industrial by- product of Al production containing mainly iron and aluminium oxides) were employed. The corrosive environment, where the tests were made, was a 3.5% NaCl static solution maintained at 30 °C by circulating thermostats. Steel coupons of the dimensions 50 x 60 x 0.1 mm were prepared. The corrosion products were mechanically removed from the surface of the steel specimens and also by immersion in inhibited hydrochloric acid for 15 min. They were washed by deionised water, dried, marked and weighed.

Two procedures were followed:

- Immersion of 10 specimens of the different types [uncoated or coated with the protective systems (a 90μm dry film thickness)] for 2 months at 30 °C, in the 3.5% NaCl solution. At the end of the exposure, the coatings were removed first by the epoxy diluent and then mechanically (using an acrylic knife), were immersed in inhibited hydrochloric acid to eliminate the corrosion products formed during the exposure, washed with deionised water, dried and weighed to obtain the weight loss.
- Immersion of 20 specimens of each type in the 3.5% NaCl solution and at several time intervals, 5 specimens of each type were removed and treated as above to measure the weight change as a function of time.

With these two procedures the following measurements were taken in order to prove the validity of the idea of using n-semiconductor pigments for the protection of steel.

1. Weight loss of steel specimens uncoated and coated with E.R. unpigmented and pigmented with the n-semiconductors (first procedure), after exposure in 3.5% NaCl solution for 2 months (30 °C):
   Comparison between different semiconductive pigments

2. Weight loss vs. time (second procedure) in 3.5% NaCl solution (30 °C) of steel specimens uncoated, coated with E.R., unpigmented and pigmented with two different types of ZnO and with Zn powder:
   Comparison between the same oxide with different semiconductive properties

3. Weight loss vs. time (3.5% NaCl, 30 °C) of steel specimens uncoated, coated with unpigmented E. R. and pigmented with n-semiconductor Al_2O_3 plain and doped with MgO:
   Comparison between plain and doped semiconductors

4. Weight loss vs. time (3.5% NaCl, 30 °C) of steel specimens uncoated, coated with unpigmented C. R. and C.T.E. and pigmented with n-semiconductor Al_2O_3:
   Comparison of the anticorrosive properties of known paints plain and those pigmented with n-semiconductor Al_2O_3.

5. As 4 under cathodic protection (C.P.) of -850 mV:
   Influence of the cathodic protection on the protective systems

6. Weight loss (3.5% NaCl, 30 °C, 30 days) of steel specimens coated with unpigmented E.R. and pigmented with n-semiconductor vs. cathodic protection voltage:
   Energy save using n-semiconductor pigments

7. Weight loss vs. time of concrete rebars (reinforced concrete specimens in 3.5% NaCl, at 30 °C, for 8 months with and without different percentage of red mud):
   Influence of semiconductor on the corrosion of concrete rebars

2.2 First Measurement Series (Table 1)

Table 1 *Comparison between different n-semiconductor pigments 30% vs.E.R. Weight loss of steel specimens in 3.5% NaCl solution for an exposure of 2 months at 30 °C. (Mean of 10 specimens for each case, error ± 3%)*

Type of Pigment	% Protection
Fe_3O_4	~90
Al_2O_3	~81
ZnO	~75
Epoxy resin	~58
Fe_2O_3	~50
Uncoated	0

2.3 Second Measurements Series (Figure 1)

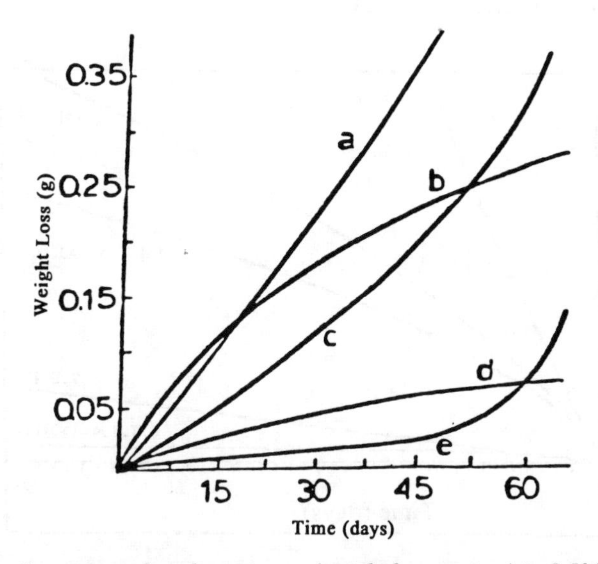

Figure 1 *Corrosion of steel specimens (weight loss vs.time) in 3.5% NaCl solution at 30 °C, a: uncoated metal, b: 30% ZnO pigmented epoxy, c: epoxy coating, d: 30% ZnO with ameliorated n-semiconductive properties, e: 30% Zn powdcr pigmented epoxy:Comparison between ZnO prep. by different methods (b, d).*

2.4 Third Measurement Series (Figure 2)

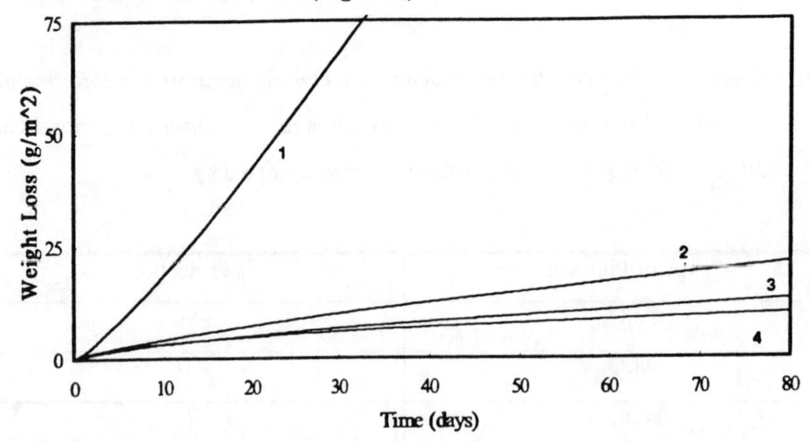

Figure 2 *Corrosion (weight loss vs. time) of uncoated steel specimens (1), coated with E.R.. (2), with E.R.. pigmented with n-semiconductor Al_2O_3 (3) and n-semiconductor Al_2O_3 doped with MgO (4): Comparison between plain and doped n-semiconductor (3,4).*

2.5 Fourth measurement series (Figure 3)

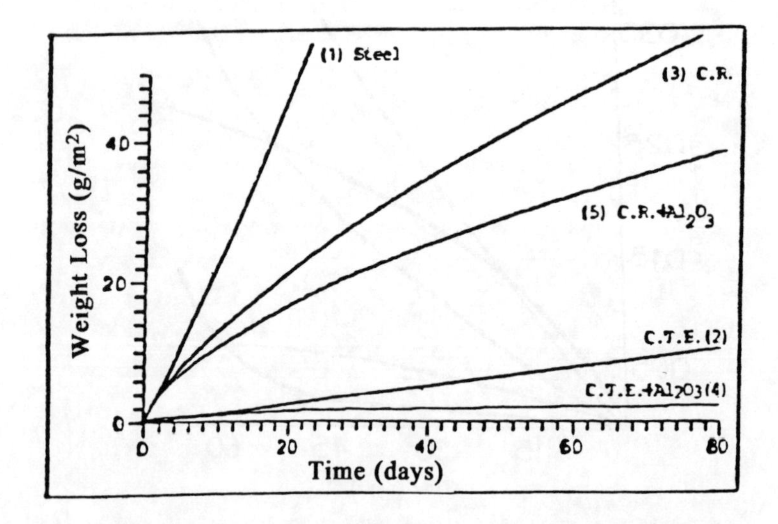

Figure 3 *Corrosion (weight loss vs. time in 3.5% NaCl, 30 °C) of steel specimens uncoated (1) and coated with plain C.R.. (3) and C.T.E. (2) and pigmented with n-semiconductor Al_2O_3 (5 and 4 respectively): Comparison of known anticorrosive paints plain and pigmented with n-semiconductor.*

2.6 Fifth Measurement Series (Figure 4)

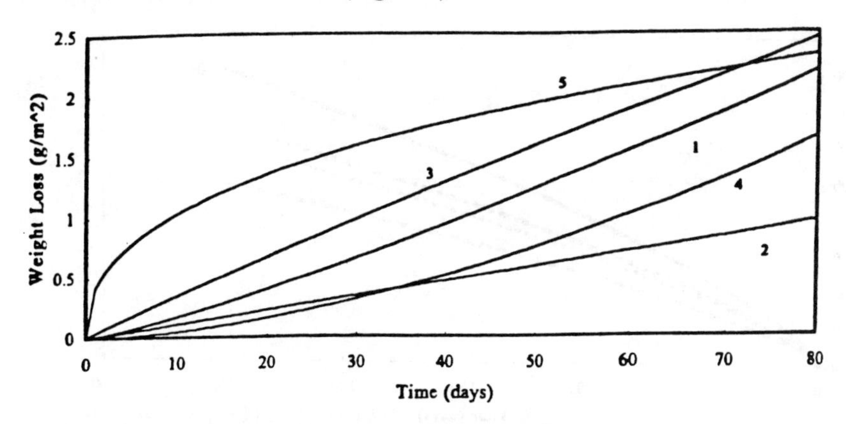

Figure 4 *Corrosion (weight loss vs. time in 3.5% NaCl, 30 °C) of the same specimens as in Figure 3: 1. uncoatcd steel, 2. coated with C.T.E., 3. coated with C.R.., 4. coated with C.T.E.+ n-semiconductor* Al_2O_3, *5. coated with C.R..+ n-semiconductor* Al_2O_3 *under cathodic protection of -850 mV: Influence of the cathodic protection on the protective systems (to be compared with Figure 3).*

2.7 Sixth Measurement Series (Figure 5)

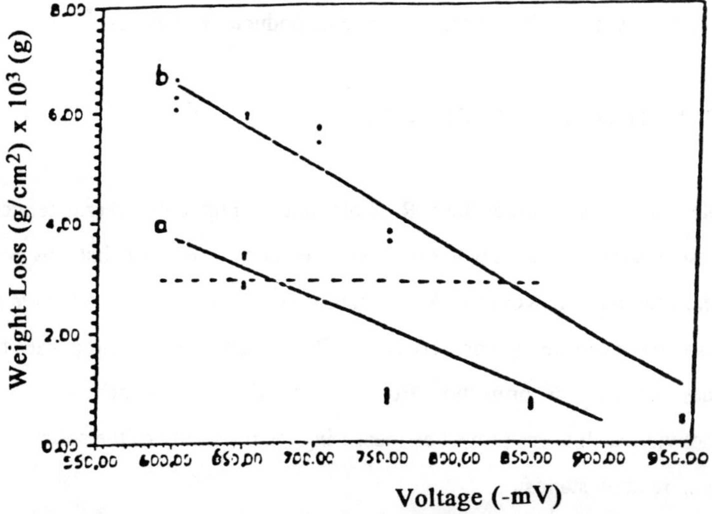

Figure 5 *Corrosion (weight loss: 3.5% NaCl, 30 °C, 30 days) of steel specimens coated with unpigmented C.R. (b) and pigmented with n-semiconductor with lower semiconductor intensity (red mud): Energy save*

2.8 Seventh Measurement Series (Figure 6)

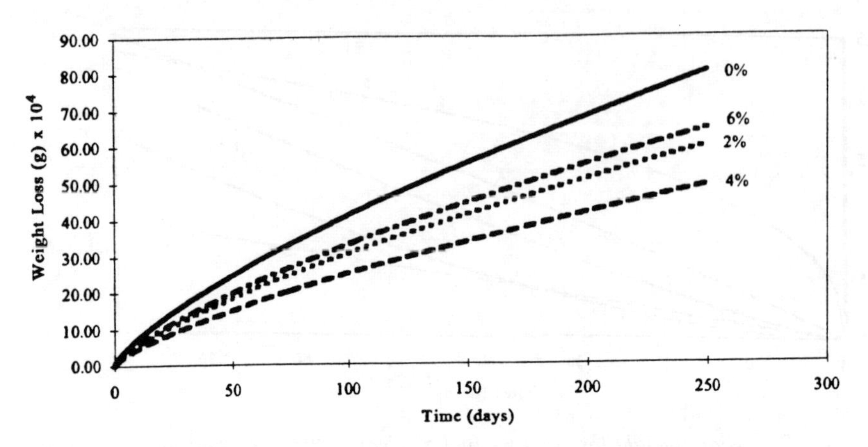

Figure 6 *Corrosion (weight loss of steel rebars vs. % R.M. additive in concrete, after exposure in 3.5% NaCl solution, 30 °C) for different percentages.*

Besides, similar results are acquired by Red Mud on the protection of rebars when n-semiconductors in the form of additives are added in the concrete, replacing 4% of the sand, such as Terra Theraic (a natural material of volcanic origin commonly used as concrete additive) with randomly acquired n-semiconductor properties[7,8,9].

3. OBSERVATIONS AND DISCUSSION

From Table 1 it is obvious that the E.R. containing as pigments several semiconductors protect the steel substrate much better than unpigmented. The Fe_2O_3 that was not prepared by a special method that would make it acquire pronounced semiconductive properties does not increase the protective properties of E.R. The anticorrosive properties of the four other oxides increase the protection from ZnO to Fe_3O_4 with increasing intensity of semiconductivity, as this was proved by measuring the resistivity of the solid oxides where the following relation stands:

$$R_{ZnO} \rangle R_{Al_2O_3} \rangle R_{Fe_3O_4} \qquad (1)$$

In Figure 1 we see once more that the E.R. pigmented with ZnO (b and d) protects steel better than the unpigmented (c). The ZnO (d) was specifically prepared to possess higher n-semiconductive properties: Higher conductivity. The E.R. pigmented with 30% Zn powder [less than for Zinc Rich Epoxy (90%)] protects better even than E.R. with ZnO (d) pigmented, but after a time interval the Zn-powder is sacrified for the protection and the corrosion increases more than the one imposed by the ZnO (d).

In Figure 2 the influence of the n-semiconductivity on the protective properties of E.R. is once more confirmed (compare curves 2 and 3) with the additional observation that, if the n-semiconductivity is reinforced by doping the n-semiconductor Al_2O_3, the protective properties increase.

In addition to the above mentioned, the pigmentation of the two known anticorrosive paints C.R. and C.T.E with the n-semiconductor Al_2O_3 follows, to an appreciable extent, the decrease in the corrosion of steel, (Figure 3).

If besides the addition of n-semiconductor Al_2O_3 the protective systems are reinforced by cathodic protection (Figure 4), steel is seriously protected (compare 1 in Figure 3 and 4) the protection by C.T.E increases, but the difference with uncoated steel is not so much as without Cathodic Protection (compare 1 and 2 in Figure 3 with 1 and 2 in Figure 4) without arriving at an overvoltage limit. On the contrary in the presence of C.P., steel corroded more if coated by C.R. pigmented or not with n-semiconductor Al_2O_3. This is clearly an overvoltage phenomenon which confirms the way the n-semiconductor acts i.e. as a type of C.P. overlapping the impressed current. Thus if we want to also use C.P. with C.R. we must apply less than -850 mV impressed current or a lower content of semiconductor in the E.R. or a semiconductor with lower semiconductive properties. The last is confirmed by Figure 5, where Red Mud was used. R.M. is a by product of the aluminium industry and contains principally Fe_2O_3 and Al_2O_3 that have randomly acquired n-semiconductor properties. We see that in this case if we use C.P. we increase the protective properties of pigmented C.R. When we compare now for the same corrosion (same weight loss) the two curves with and without semiconductor we see that the same result in protection is acquired if we use -850 mV C.P. for non pigmented C.R. and only -680 mV for pigmented C.R., i.e. we save much energy during the application of C.P.

We further see in Figure 6 that protection of concrete rebars can be acquired by replacing 4% of sand by R.M. or Terra Theraic, that both contain $Al_2O_3$and Fe_2O_3 with randomly acquired n-semiconductor properties during their production: dissolution of Al_2O_3 of bauxite by the Bayer method under high pressures and temperature and filtration for R.M., sudden cooling of material, being of volcanic origin for Terra Theraic.

4. CONCLUSIONS

1. The protection of steel by externally imposed negative charge (cathodic protection or by sacrificial anodes etc.) can also be applied by polymeric vehicles or by concrete both pigmented with n-semiconductors, having the predisposition to offer electrons.

2. The n-semiconductivity of substances depends on their structure (excess of cations or lack of anions in the crystal stoichiometry) which highly depends on the method of their preparation and precisely on their doping.

3. Except for the artificial preparation of n-semiconductors (Al_2O_3, Fe_3O_4, ZnO and doped), some natural products, such as Terra Theraic (of volcanic origin), and industrial by-products such as Red Mud (by-product of the Al_2O_3 industry) possess n-semiconductive properties, due to randomly acquired n-semiconductive properties on the Fe_2O_3 and Al_2O_3 that are contained within.

4. The anticorrosive properties of n-semiconductors are more pronounced if they are doped by MgO. These semiconductors do not get exhausted as materials or electron donors as known from their application as catalysts in the industry.

5. As pigments in polymer vehicle or in known anticorrosive paints they satisfactorily protect steel and in the latter case, increase the anticorrosive properties of the known paints Coal Tar Epoxy and Chlorinated Rubber.

6. As additives in concrete they protect the steel reinforcements.

7. When C.P. and pigmented anticorrosive paints are simultaneously employed the over voltage phenomenon appears.

8. When cathodic protection and pigmented anticorrosive paints with low content of semiconductor or low voltage are simultaneously employed, the semiconductor pigments in the paints replace part of the cathodic voltage that follow to save energy.

9. As sensitisers they protect the polymeric vehicle from UV attack.

10. The same systems can also protect marbles exposed in a hostile urban environment[10,16], because the sulphation has a similar mechanism to the uniform corrosion of metals[17-21].

References

1. Th. Skoulikidis and A. Tsakopoulos, *Br.Corros. J.,* 1978, **13,** 130.
2. Th. Skoulikidis, A. Moropoulou and A. Tsakopoulos, Proc. 5th Int. Congress on Marine Corrosion and Fouling, Barcelona, 1980, 541.
3. Th. Skoulikidis, A.Tsakopoulos and A..Moropoulou, Int.Symposium on the Corrosion Effect of Stray Currents and the Techniques for Evaluating Corrosion of Rebars in Concrete (ASTM), Williamsburg, Virginia, 1984, ASTM Special Publ., 1986, **906,** 15.
4. Th. Skoulikidis and S.Vlachos, 12th Proc.Scandinavian Corrosion Congress & EUROCORR'92, Espoo, Finland, 1992, **1**, 453.
5. Th. Skoulikidis and P. Vassillou, 6th Int. Cong. on Marine Corrosion and Fouling, Athens 6/1984, *Corrosion,* 335.
6. Th. Skoulikidis and P.Vassiliou, 3rd Int.Cong. on Marine Technology, Athens, 1984, 563.
7. Th. Skoulikidis, P.Vassiliou and N.Diamantis, Proc. 5th Int. Congress on Marine Technology, Athens, 1990, 155.

8. Th. Skoulikidis, P. Vassiliou and N. Diamantis, Proc. 12th Scandinavian Corrosion Congress & EUROCORR' 92, Espoo, Finland, 1992, **II.**, 475.
9. Th. Skoulikidis, *Bulletin de Liaison du COIPM,* 1986, **19,** 33.
10. Th.Skoulikidis, D.Charalambous and E.Papakonstantinou, Proc.Int.Symposium on Engineering Geology as Related to the Study, Preservation and Protection of Ancient Works, Monuments and Historical Sites, Athens, 1988, 871
11. Th. Skoulikidis, D.Charalambous and E.Kalifatidou, Proc.VIth Int.Congress on Deterioration and Conservation of Stone, Torun (Poland), 1988, 535.
12. Th. Skoulikidis and E.Kritikou, STREMA Int. Conference, Seville, 1991.
13. Th. Skoulikidis and E.Kritikou, Proc. 2nd Int. Symposium for the Conservation of Monuments in the Mediterranian Basin, Geneva, 1991, 389.
14. Th. Skoulikidis and E.Kritikou, Proc. 7th Int.Congress on Deterioration and Conservation of Stones, Lisbon, 1992, **III**, 1137.
15. Th. Skoulikidis and E.Kritikou, Proc. 3rd Int.Conference STREMA 93, Bath (U.K.), 1993, 241.
16. Th. Skoulikidis and D.Charalambous, *Br. Corros. J,* 1981, **16,** 70.
17. Th. Skoulikidis, D.Charalambous and P.Papakonstantinou-Ziotis, Proc.4th Int. Congress on the Deterioration and Protection of Building Stones, Louisville, 1992, 307.
18. Th. Skoulikidis, D.Charalambous and P.Papakonstantinou-Ziotis, *Br.Corros. J,* 1983 **18,** 200.
19. Th. Skoulikidis and D.Charalambous, Proc.VIth Int. Congress on the Deterioration and Preservation of Stones, Lausanne, 1985, 547
20. Th. Skoulikidis, D.Charalambous and M.Kyrkos, Proc. Conference on the Recent Advances in the Conservation and Analysis of Artifacts, University of London, London, 1987, 383.
21. Th. Skoulikidis and A.Ragoussis, *Corrosion,* 1992, **48,** 666.

10. QUANTUM CHEMICAL STUDY OF NITROGEN AND SULPHUR CONTAINING SUBSTANCES AS INHIBITORS OF THE CORROSION AND HYDROGENATION OF STEEL

G.S.Beloglazov, S.M.Beloglazov

Perm State University, Perm, Russia, Kaliningrad State University, Dept. Chem.
14 Alexander Nevskiul, Kaliningrad 236041 Russia. 0112-338870

1. INTRODUCTION

The advantages of using of organic corrosion inhibitors for metal protection in the different natural and industrial media have been known for a considerable time [1]. Different organic substances are used as steel corrosion inhibitors in acid and water/ salt media[1-6], but at the present time their choice is purely empirical. An understanding of the basic relationship between the chemical structure of organic molecules and their corresponding inhibitive efficiency would enable a selective, and in some cases, low cost synthesis of highly efficient inhibitors to combat steel corrosion and hydrogen absorption[6,7]. In general the most effective inhibitors include compounds with amine, or substituted amine, groups[7,8]. Our initial comprehensive data on the inhibitors of hydrogen absorption of steel (so called "hydrogenation of steel") by electrochemical corrosion or metal deposition, as well as by cathodic protection from marine corrosion and technological media[7,9], revealed that derivatives of amines and heterocyclic N-containing compounds were very effective agents for such processes.

In 1971 several attempts were made to apply quantum chemical calculations to this problem [10,11] and progress has been made in using such an approach to a number of molecules belonging to several groups of N-containing organic compounds [11-15]. In this paper we propose a method for determining the specific relationship between structure and inhibitor efficiency for N- and S-containing substances against the corrosion of mild steel in seawater infected with sulphate reducing bacteria (SRB). Unfortunately the important task of the quantitative estimation of structure-property relationships cannot be solved in a general way because each individual class of chemical compounds has its own internal regularities. In this article, we consider derivatives of substituted amines possessing a diethylamine group which have been demonstrated to be effective corrosion protection agents for steel in various aqueous environments. Severe experimental conditions were produced by the use of aqueous solutions containing sulphate reducing bacteria.

2. EXPERIMENTAL DETAILS

Several methods for the investigation into the corrosion of steel in SRB infected seawater were used to obtain parameters characterising the processes of:

(i) steel corrosion

(ii) adsorption of hydrogen on local cathodes

(iii) SRB population growth in different corrosion media (variations in types and concentrations of organic substances with a known degree of inhibition).

2.1 Corrosion media used in experiments.

In contrast with earlier investigations [7-9], much of the present study was carried out in aqueous media deteriorated with SRB. The experiments were performed with either artificial seawater or a solution of inorganic salts in water (standardised microbiological Postgate medium B[16]. Artificial seawater consisted of sodium chloride 7.5 g dm^{-3}, magnesium sulphate 1.0 g dm^{-3}, sodium carbonate 1.0 g dm^{-3}, sodium sulphate 2.0 g dm^{-3}, sodium dihydrogen phosphate 0. 5 g dm^{-3} and calcium lactate 2.0 g dm^{-3}. A pH of 6.8-7.0 was measured for this corrosion medium at the start of the experiment. Corrosive media were sterilised by boiling for 2 hours before the experiments.

In order to develop our ideas concerning structure-property relationships in organic compounds with regard to the function of the corrosion inhibitor, hydrogen absorption inhibitor and biocide against SRB, several organic compounds of the general formula R-$N(C_2H_5)_2$ were synthesised and investigated in SRB containing corrosion media.

SRB are found world wide in many industrial and natural environments e.g. in soil, sea and freshwater, wastewater and petroleum products etc.[17-20]. Water fuel tanks, paper mills, heat exchanger-closed water systems and wastewater treatment tanks are liable to corrosion by the action of these bacteria. The corrosion activity of SRB has been explained on the basis of their hydrogenase activity within the biofilm and sulphate reductase activity in the bulk of solution[21]. SRB can use atomic or molecular hydrogen or protons present at the cathode in the formation of hydrogen sulphide via the action of hydrogenase. The removal of H or H_2 from the cathode surface results in a cathodic depolarisation and promotes metallic corrosion. Hydrogen sulphide acts as a strong stimulant for hydrogen absorption (HA) by steel subsurface layers in the corrosion process[9,22]. SRB were isolated from sediments from the Vonychka river by repeated sub-culturing in Postgate medium B. The reference series of experiments were carried out using pure cultures of *Desulphosarcina variabilis* and *Desulphovibrio baculatus* from the Institute of Physiology and Biochemistry of Microbes (Pushchino). "Parallel studies" showed that isolates of a wild culture were more active than those from a reference (museum) culture.

2.2 Metal specimens

Mild steel (carbon <0. 2 %) specimens, 50 x 20 x 1.5 mm, were cut from sheet steel, polished with fine grained carborundum paper and cleaned with lime paste to remove grease from the surface. Rinsing in tap and distilled water was followed by UV sterilisation for 20 minutes on each side. After sterilisation, weighed steel specimens were placed in the test tubes (suspended from a mild steel wire in order to measure the electrode potential) containing 50 cm^3 of the corrosion medium. Two cm^3 of a 2-day old bacterial culture were introduced under sterile conditions. The test tubes were hermetically sealed with sterilised plugs and stored at 310 K.

2.3 Experimental techniques

The numbers of bacterial cells were counted every 24 hours using a microscope equipped with a *Goryayew* camera and a phase contrast addition FK-4. The redox potential, E_h, and pH values were measured using a Radelskis OP210/3 microanalyser. Electrode potentials were measured against an Ag/AgCl reference electrode.

It was estimated that a 48 hour period was sufficient for the optimal development of SRB under the experimental conditions. The organic species were therefore added after this period to test their inhibiting and biocidal properties against the corrosion and hydrogenation of steel. After 8 days the specimens were removed from the corrosive medium, the black sediment of iron sulphides washed away and the specimens dried and weighed.

The quantity of hydrogen absorbed by the steel was estimated by measuring of decreased concentration of oxygen dissolved in the anolyte[23]. The oxygen interacted with the hydrogen resulting from the dissolution of the steel specimen on anodic polarisation. A large Pt plate in the anode compartment acted as a catalyst for this interaction.

At the same time as the inhibitor was added to the test tubes, steel wire specimens 0.5 mm in diameter were placed in some of the test tubes. After 8 days of exposure the wire specimens were taken out and washed and the plasticity loss due to hydrogen absorption by the steel during corrosion determined by the number of twisting of axial rotations of wire fixed in clamps of machine K5.

3. QUANTUM CHEMICAL COMPUTATIONS

The experimentally measured inhibitor efficiency of organic compounds with diethylamine groups was correlated with the results of quantum chemical MNDO computations. Mulliken net charges and their other combinations, energies of boundary molecular orbitals (HOMO and LUMO) as well as their sum and differences were determined. Quantum chemical computations were made for organic molecules in the free state and not for their clusters with one or more surface iron atoms, because the particular quantum characteristics of an iron atom are not currently available.

4. RESULTS

In the reference solutions where no inhibitor was added, the redox potential E_h of the corrosion medium changed slightly to more negative values during the first few days, then shifted in the positive direction (Figure 1), usually reaching values of –230 to -250 mV.

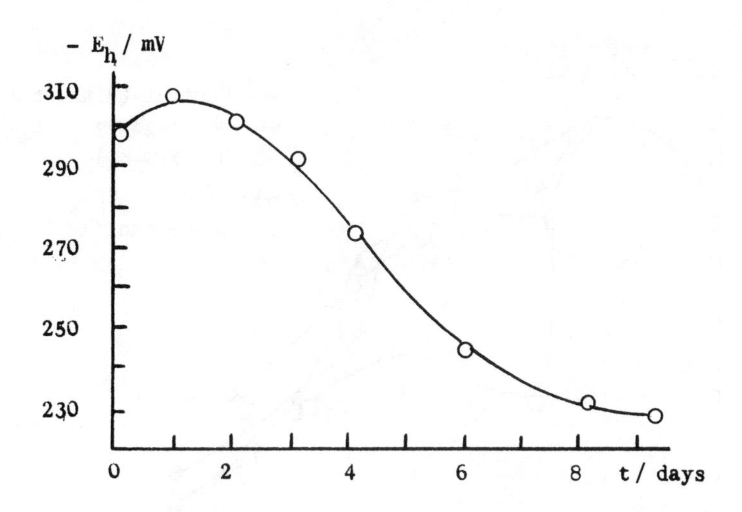

Figure 1 *Variation in the value of the redox potential of a water-salt corrosion mediumcontaining SRB*

The acidity of the medium decreases during the first 8 days of exposure, see Figure 2. The number of SRB cells increases up to 8 x 10^7 cm^{-3} during the first 48 hours and then steadily decreases, as can be seen in Figure 3.

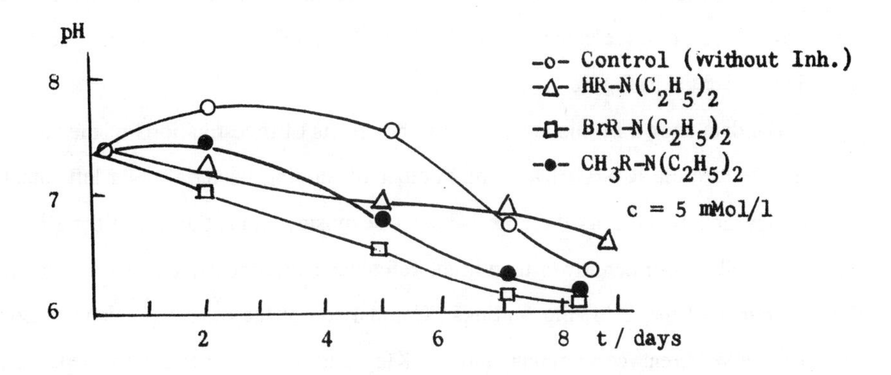

Figure 2 *Variation in the value of the pH of a water-salt media containing inhibitors and SRB with exposure time*

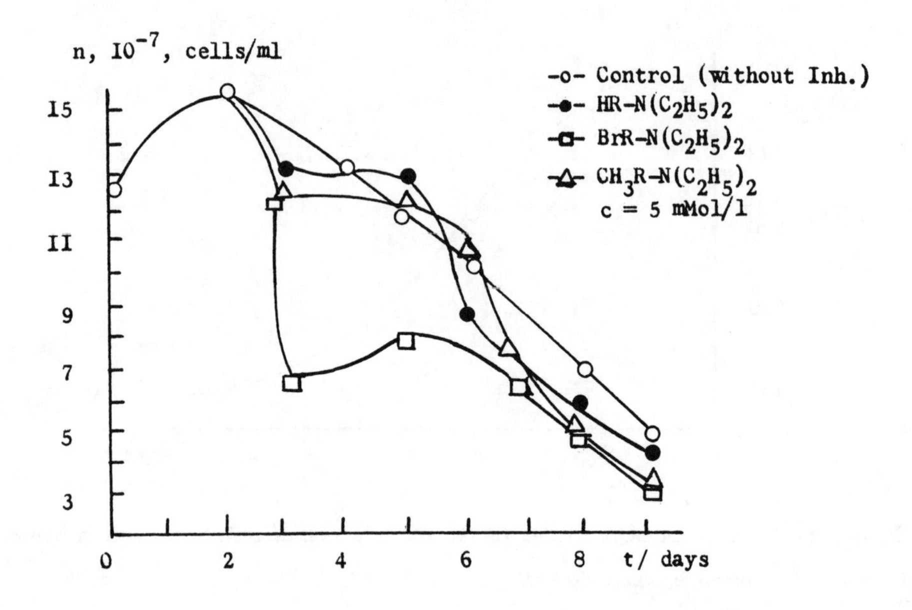

Figure 3 *Change in the SRB cell numbers with exposure time in water-salt media with additives*

The addition of organic compounds, which inhibit the corrosion of steel, to the electrolytes resulted in a rapid shift of the redox potential to more positive values followed by a slower change in the same direction. The typical effect of such organic compounds on E_h is represented in Figure 4.

The data shown in this are based on the results of investigations of the substances of a general formula $R\text{-}N(C_2H_5)_2$. The number of moving bacterial cells left intact was already significantly decreased on the 3rd day of exposure and at the end of the 8th day the number of cells was much lower than in the reference experiments. This should emphasise the importance of the following findings. Comparison of the time dependencies E(t) and n(t) for the different compounds and of K(c) and n(c) concentration dependencies, suggests the similar behaviour of $E_h(t)$, n(t) and K(t). Owing to the presence of organic substances with an inhibiting efficiency, the pH of the corrosion media significantly decreases in comparison with that of the reference experiment, Figure 2.

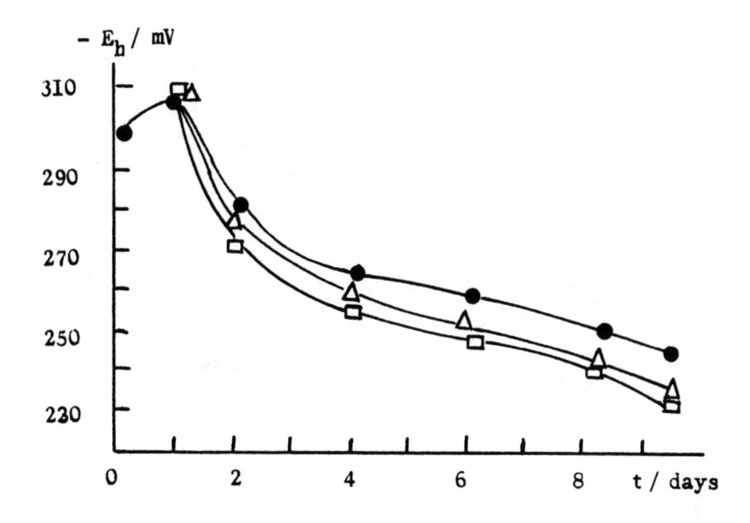

Figure 4 *The effect of organic compounds as additives to saline corrosion media on the redox potential*

The quantity of hydrogen absorbed by the steel sub-surface layers during corrosion, V_H, was less in the cases where the medium contained additional organic substances compared to the reference experimental series. Typical effects of these organic compounds on hydrogen absorption are represented in Figure 5.

The greater the corrosion inhibitor concentration in the corrosive media, the greater the suppression of hydrogen absorption by the steel. The loss of steel plasticity, measured on wire specimens, was significantly decreased by the addition of organic substances with inhibiting efficiency to the corrosion medium, Table 1. The present results give evidence of the possibility of the parallel influence of the changes in structure and content of organic molecules as efficient inhibitors of corrosion and hydrogenation of steel on the one hand, and inhibitors of SRB development on the other. The best inhibitor of steel hydrogenation during corrosion in water-salt media in the presence of SRB proved to be that with R = p-BrC_6H_4; this was also the best inhibitor of corrosion and the best biocide. The greater inhibiting efficiency of the addition of a derivative with R= p-Cl C_6H_4 in comparison with R= p-CH_3 C_6H_4 is in agreement with previous results[9] on the study of hydrogenation of carbon steel under cathodic polarisation in solutions of sulphuric acid.

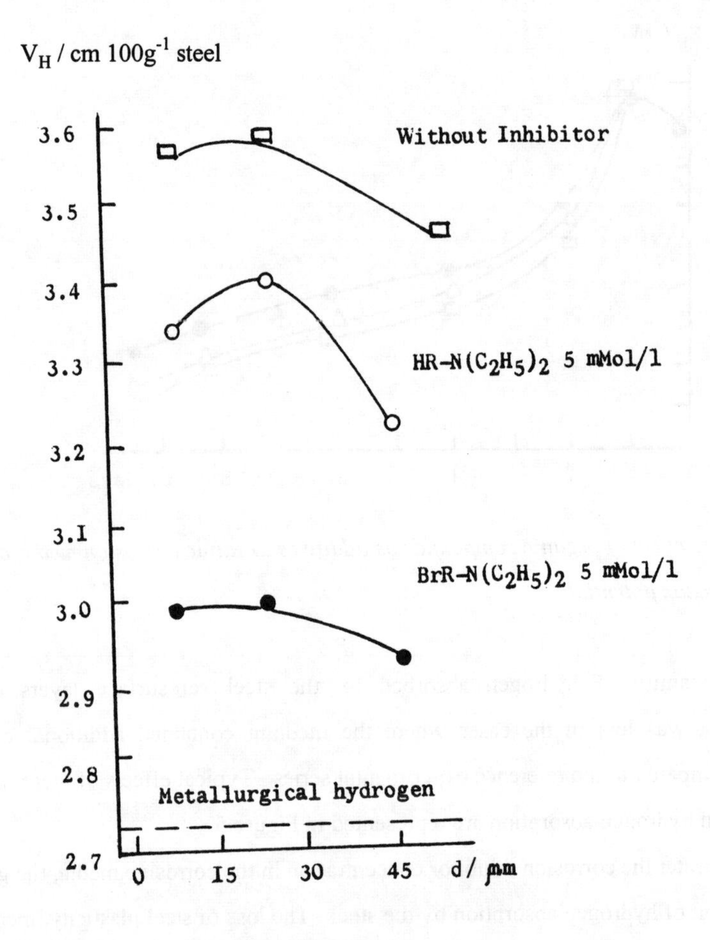

Figure 5 *Quantity of adsorbed hydrogen in steel subsurface layer as a consequence of corrosion in saline media containing SRB*

Despite the similar shape of the n(c) and K(c) plots, for all types of compounds examined[4] there was no exact correlation between their efficiencies as biocides and inhibitors of the hydrogenation of steel. It is clear that both the biocidal and corrosion suppressing activity of organic N- and S-containing compounds are based on the adsorption of these molecules onto the solid surface but their nature and composition is not equivalent. Additionally, the activity must be a function of ability of organic molecules to penetrate the biological cell membranes. For example, the detergent OP10 with a high surface activity is a very effective biocide against SRB but the least effective inhibitor for the corrosion of steel [15].

Table 1 *Relative retention Z_h of plasticity of steel wire specimens after the hydrogenation process. Cathodic polarisation current density = 0.02 A cm^{-2} for 48 minutes at pH=2. Electrolyte 1 = 0.1 M H_2SO_4, Electrolyte 2 = Electrolyte 1 + H_2SeO_3 (5 g dm^{-3}). R1 = $RCHCH_3$, R2 = $RCHCH{=}CHCH_3$, R3 = $RC(C_2H_5)_2$, R4 = $RC(CH_3)(C_3H_7)$, R5 = $RC(CH_3)(C\text{-}(CH_3)_3)$, R6 = $RC(CH_3)(CH_2\text{-}CH(CH_3)_3)$, R7 = $RC(CH_3)(C_6H_{13})$, R8 = $RCH\text{-}C_6H_4\text{-}m\text{-}OH$, R9 = R=N(4)-N(3)-[C(5)-(N(1)CCH_3-C-CH-C(6)-OH-N(2)-]*

Z_h /%	R1	R2	R3	R4	R5	R6	R7	R8	R9
Electrolyte 1	62	99	39	77	69	69	62	23	72
Electrolyte 2	62	92	46	77	77	77	54	60	77

Table 2 shows the values of some of the MNDO quantum chemical results obtained for isolated molecules and the efficiencies of the inhibitive action of the corresponding species which have been measured experimentally in various conditions: Z_s and Z_b are the corresponding values which characterise the decrease in the relative weight loss of steel samples in Postgate B media due to the inhibiting action.

As seen from Table 2, the strong correlation of Z_h and Mulliken net charges on atoms N(3), N(4), C(5) and C(6) and the HOMO energy suggests a different type of interaction between the active centres/electronic parameters and the protective effect of the species against hydrogenation or corrosion (in the latter, only the charges on C(6) and N(2) significantly correlate with Z_b). The non-significant values of LUMO correlation with all Z data suggests the absence of the role of acceptor properties in inhibiting molecules (at least of the class having been investigated) in all processes except the case of corrosion in sterile media (Zs).

Table 2 *Correlation between MNDO quantum chemical results and inhibiting efficiencies Z of the corresponding species for various corrosion/hydrogenation conditions. "s" and "b" denote sterile and Desulphvibrio desulphuricans infected media respectively.*

Xi, Quantum chemical characteristics (MNDO)	Correlation of x_i with Z_h, %	Correlation of x_I with Z_s. %	Correlation of x_i with Z_b, %
HOMO energy	-32	-93	-43
LUMO energy	8	90	37
ΔE HOMO-LUMO	25	92	41
Net N(1)	20	93	44
Atomic N(2)	-40	-98	-57
Charges N(3)	-43	93	44
Charges N(4)	-33	-92	-40
Charges C(5)	-16	93	44
Charges C(6)	50	53	97
Sum of atomic charges on CH group	-28	-83	-23

5. CONCLUSIONS

The derivatives of substituted amines with a diethylamine group are effective inhibitors of the corrosion and hydrogenation of steel by corrosion in saline media contaminated with SRB. The results give evidence of the possibility of the parallel influence of the changes in structure and content of the organic molecules on the inhibiting properties and development of SRB. It appears that different active centres of inhibiting molecules of organic N- and S-containing compounds are responsible for their protective action against the absorption of hydrogen by the steel sub-surface layers by corrosion and corrosion process depending also on the presence of SRB.

References

1. I.N. Putilova, S.A. Balesin and V.P.Barannik, "Inhibitors of Metal Corrosion", Goshimizdat, Moscow, 1958.
2. Z.I. Levin, A.1. Altsybeyeva, " Inhibitors of corrosion", Leningrad, 1968.
3. I.L. Rosenfeld, " Inhibitors of Corrosion", Moscow, 1968.
4. J.Bregman, " Inhibitors of Corrosion", Moscow, 1966.
5. J. S. Robinson, "Corrosion Inhibitors. Recent Developments". Noyes Data Corporation, Park Ridge, 1979.
6. L.I Antropov, E.M. Makushin and V.F. Panasenko, "Metal Corrosion Inhibitors".
7. S.M. Beloglasov, *'Scientifc Mem. of Perm Univ.'*, 1968, No. 1946, 40.
8. S.M. Beloglazov, Thesis, Perm. 1960.
9. S.M. Beloglazov, "Hydrogenation of Steel during Electrochemical Processes", Leningrad University Press, Leningrad, 1975.
10. J. Vosta and J. Eliasek, *Corros. Sci.,* 1971, **11**, 223.
11. G. S. Beloglazov, *"Metal Corrosion and Prevention* ", (Kaliningrad),1983, **6**, 14.
12. S. M. Beloglazov and G. S. Beloglazov, Proc. Eurocorr 91, Budapest, 1991, p. 134.
13. G.S. Beloglazov, S.M. Beloglazov and O.D.Gladysh, " Progress in the Understanding and Prevention of Corrosion", The Institute of Materials, London, 1993, Vol. 2, p 900.
14. S.M. Beloglazov, G. S. Beloglazov and S.N.Uss, *ibid,* p 1185.
15. S.M.Beloglasov, T.B. Postnikova and E.V. Frolowa, *Phys. Chem.Mechanics of Materials (USSR),*1986, **22**, 108.
16. E.I.Andreyuk, I.A.Kozlova, " Litotrofnye bakterii i Mikrobiologicheskaya korrosiya", Naukova Dumka, Kiev, 1980.
17. C.R. Gatellier, *Corrosion Traitements, Prot.,*1973, **21**, 103.
18. G.Kobrin, *Materials Perform.,* 1976, **15**, No 7, 38.

19. I.Heyer and W. Schwartz, Z *allg.Mikrobiologie,* 1970, **10**, 545.

20. R.E. Tatnall, *Materials Perform.*, 1984, **23**, 13.

21. I.M. Sharpleg, *Corrosion,* 1961, **17**, 386.

22. Z.A.Iofa, *Mater.Protect*, 1974, **10**, 300.

23. Y.A.Klyachko, I.Y.Shklovskaya and I.A.Ivanova, Zavodskaya Laboratories, No.9, 1970, 1090.

11. IMPROVED SACRIFICIAL ANODES FOR THE PROTECTION OF OFFSHORE STRUCTURES

P.L Bonora.*, S. Rossi*, L. Benedetti* and M. Draghetti†

* *Laboratory of Electrochemistry, Materials Engineering Department, University of Trento, Mesiano (TN) Italy 38050*

† *AGIP Offshore, S.Donato (MI) Italy 20097*

I. INTRODUCTION

Sacrificial anode protection systems are dimensioned according to the required protective current, so that the entire structure to be protected is polarised to a value sufficient to render corrosion negligible. The required current density is a function of the surface state of the structure, the sea conditions, the electrochemical properties of the anode alloy and of anode geometry.

The surface state of an immersed cathodically protected structure is strongly influenced by the presence of any calcareous deposits which precipitate as a result of polarisation. The effect of such a deposit is to limit current circulation in the galvanic element which the short circuited anodes and structure create. Hence, the maximum protection current is needed when the structure is new and no deposits are present. The current later supplied is also a function of the deposits formed initially. The anodic system must in any case be dimensioned to supply the initial current required, and must guarantee protection over the entire foreseen lifetime of the structure.

The development of a new sacrificial anode for the protection of offshore structures against corrosion is decribed in this chapter. This anode has a composite structure in which an anodic material composed of an aluminium alloy forms the covering on the internal support for the anode and a magnesium alloy, with a thickness of approx. 1.5 cm, forms an external skin.

Thanks to its special composite structure, this new anode improves the level of protection on the surfaces of offshore structures through increased formation of compact protective deposits. It also offers reductions in the volume and weight of the anodic material required, and, consequently, a sensible reduction in costs[1]. In these composite anodes, the surface magnesium provides the high initial currents needed, without oversizing. Later, after complete dissolution of the magnesium, the aluminium offers cathodic protection for the remaining life of the structure, which has already been polarised and covered with a calcareous layer. The initial polarisation produced by magnesium gives a more compact surface encrustation of the protected surfaces than that obtained with aluminium anodes. Thus, to use a magnesium skin on aluminium anodes has the dual effect of allowing a rapid polarisation, sustained by single anodes over larger areas, and of producing more compact deposits which subsequently limit the current needs and, hence, the mass of aluminium required. The smaller size of such double anodes on a platform reduces the loads due to waves and currents to the platform structure, as well as reducing the material quantity (by an estimated 30 %).

2. EXPERIMENTAL

The performance of our new anode was initially verified with laboratory scale tests but was also evaluated on offshore structures provided by AGIP for this study. The laboratory tests were set up as follows. Artificial sea-water, ASTM D 1141-90, was flowed over the immersed surfaces at a temperature of 14 °C. A pH of 7.9 was maintained during the tests by addition of HCl; the bicarbonate content was held constant by daily additions of sodium bicarbonate. For short circuit galvanic testing, the Al and Mg alloys, derived from the composite anodes, were used as the anodes and the construction steel plates as the cathodes. The anode/cathode surface ratio was that used on the planing of Daria A platform i.e. 2.8% Mg and 2.4% Al. Cathodic protection potentials, protection currents per unit cathode surface and the dissolution morphology of the anodes was monitored. Two data loggers were used, AMEL Model 668/RM, and potentials were measured against Ag/AgCl reference electrodes.

The morphology of the magnesium and aluminium alloys and of the interface between these alloys was studied by optical and scanning electron microscopy (SEM), and analysed by *energy dispersive* X-ray spectroscopy (EDX) and X-ray diffraction using an X Rigaku diffractometer Model RAD B 3°.

Anodic polarisation curves were obtained in synthetic sea water, ASTM D 1141-90, with Ag/AgCl and platinum as the reference and counter electrodes respectively. A scan rate of 0.2 mV s^{-1} was used throughout.

3. RESULTS AND DISCUSSION

Anodic polarisation curves were obtained for samples of the magnesium, aluminium alloy and interface materials; the cathodic branch on the steel material used in testing was also traced. The electrochemical behaviour of the two alloys which make up the anode, with anodic/cathodic area ratios of 2.8% for the Mg anode and 2.4% for the Al anode, is well described via the anodic polarisation curves in artificial seawater (Figure 1). It is immediately apparent that the magnesium anode is able to provide a much higher density of protection current than the anodes in the aluminium alloy.

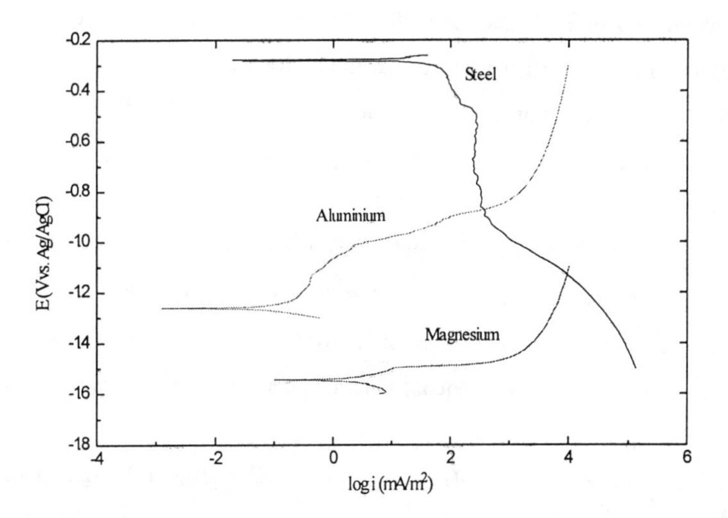

Figure 1 *Anodic polarisation curves of Al and Mg alloys and the cathodic polarisation plot for steel*

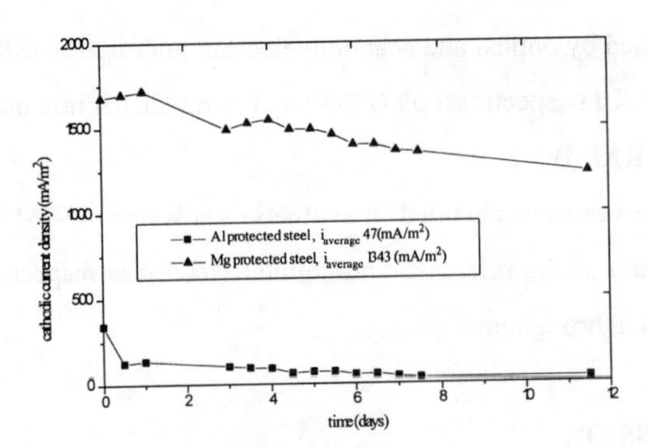

Figure 2 *Average density of cathodic current with time*

current density is observed. This effect is not as would be expected on observing the thick deposits that form on the cathode in short time (Figure 3). The result can be explained by the abundant production of hydrogen at the cathode, which probably impedes good adherence of the deposit on the steel, causing loss of the barrier effect towards the electrolyte. The production of hydrogen and the high current densities are indicative of over-protection by the Mg anode when this is used at a surface ratio R of 2.8 %.

Short circuit tests with variable area ratios were carried out to optimise the use of Mg anodes and avoid damaging effects. Using R to represent the percentage ratio between Mg area (anode) and the steel surface protected (cathode) tests were made with R varying from 0.5% to 3%. Parameters monitored with variation of R were: electrode potentials; cathode current densities; encrustation consistency; gas volumes produced at the electrodes.

The average values of cathodic plate potentials, immersed in electrolyte, vary with R from -1180 mV for R large to -1050 mV for R = 0.5% (Figure 4). That is, from high overprotection to thermodynamic conditions under which water reduction cannot take place, as the cathodic potential is nearly equal to the reversible potential in the reaction:

$$2H_2O + 2e^- \text{------>} H_2 + 2OH^- \qquad E^o = 1050 \text{ mV Ag/AgCl.}$$

The result of this tendency is exactly a strong decrease in hydrogen production at the cathode, with the decrease of R.

Figure 3 *Deposits on steel protected by Mg anode (left) and Al anode (right)*

Macroscopic deposit analysis confirms the observations made. With R>1.5% the deposit thickness is big but the adherence with substrate is very poor; with R=0.5% many zones are not covered by the calcareous encrustation. With R = 1%, the deposit is thin, compact and very adherent. This value for R appears to be the most favourable to the formation of a good calcareous deposit and minimises the autocorrosive effect of the anodic alloy[5]. These laboratory tests included the analysis of the protective capacity and of the structure of the calcareous deposits formed on the cathodes. The same techniques were used to study deposits taken from two platforms (Garibaldi D and Daria A).

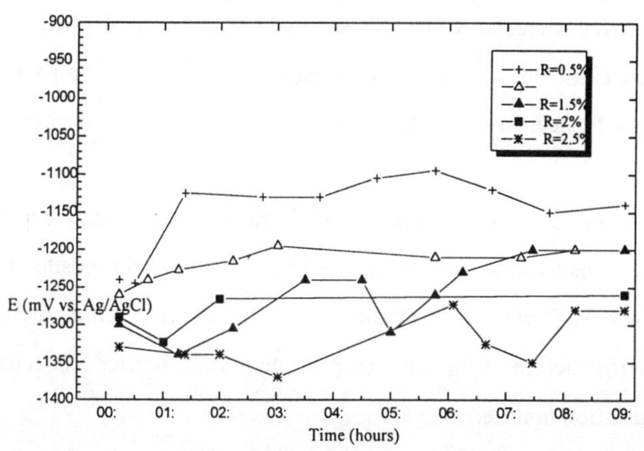

Figure 4 *Cathodic potentials on short circuit of galvanic elements with different ratio R*

From X-ray results one can note the calcareous layers close to bianode which are essentially composed of calcium carbonate was principally in the form of aragonite, instead of the calcite which constitutes the deposits on offshore structures protected by traditional anodes (far from bianode on Garibaldi platform). Thus, the magnesium ions dissolved from the anode tend to limit the deposit of calcium carbonate in the form of calcite by inhibiting its nucleation and growth, while the presence of these ions seems only to slow down the nucleation of aragonite[6,7]. X-ray diffraction studies of the deposits formed on the steel structure of the Daria A (platform located off the coast of Fano (An) in Adriatic sea totally protected with these composite anodes) revealed a significant increase in the presence of aragonite. There was also a drop in or, in many cases, the complete absence of the calcite form, and a certain quantity of magnesium hydroxide was noted on the legs of the jacket that were near the anodes. On Daria A samples the reflections due to aragonite show intensities decidedly higher than for samples from Garibaldi D. This fact suggests that aragonite formation is favoured by the low potential on the structure during protection generated by the Mg layer of the composite anodes, but that it is inhibited by other factors such as the high concentration of Mg^{2+} ions where the distance from the anode is small[6,7]. Calcite is found in all samples taken from Garibaldi D, with signals more intense in positions nearer the traditional anodes, while in samples from Daria A the signal is low and in some cases absent. The $Mg(OH)_2$ signal is strong only in samples from Daria A and in particular in those from near the composite anodes.

Scanning electron microscopy analysis shows how aragonite deposits (taken from near the bianodes on Garibaldi D) have a fine-grained regular and compact crystalline structure while deposits near the traditional anodes show a more irregular and less compact structure.

Regarding calcareous deposits formed in laboratory short-circuit experiments, it can be said that the deposit formed under the adequate action of composite anodes (R = 1%), shows calcite and Mg hydroxide in traces only if not absent, while aragonite which characterises the microstructure of the deposit guarantees good barrier characteristics and stability of the encrustation in time.

Adhesion at the interface between the two alloys is good in many areas, even though in others it is lacking due to a thin layer of alumina probably formed on the aluminium during casting and cooling. A hard fragile constituent can also be seen between the two alloys, caused by remelt of the Al during Mg casting. However the adhesion and hence electrical contact between the two alloys during dissolution of the magnesium in the service life is guaranteed by the geometry, as the magnesium covers the aluminium.

4. CONCLUSIONS

In conclusion, we can so say that the new type of sacrificial anode has been shown to guarantee the high protection currents needed, due to the absence of encrustation on cathodic areas, when the platform is new. After this period, the aluminium heart of the anode guarantees protection for the remaining life.

An initial polarisation of the protected structure by a Mg alloy allows the deposit to form better barrier properties than that produced under the action of aluminium. This is due to an increase, in the deposit, of the aragonite content with respect to calcite, another form of calcium carbonate that has lower protective properties.

Tests carried out allowed composite anode performance optimisation, and indicated that considerable advantages are available when using a surface ratio (Mg anode/cathode) of 1%.

The polarisation period due to the magnesium, lasting approximately one year, produces a successive reduction in current demand. This over a period presumably of 20 years, the life of the platform. The current reduction leads to a considerable reduction in the mass of aluminium needed, as against a thin layer of magnesium, with large economic savings given the long period.

References

I . P.L. Bonora, M. Draghetti, G.L. Valla, 'International patent'.

2. P.L. Bonora, M. Draghetti, C. Columbrita, A. Ghisetti, G.L. Valla, 'Proc. U.K.Corrosion '93', London, October, 1993, vol. 1.

3. S. Rossi, P.L. Bonora, R. Pasinetti, L. Benedetti, M. Draghetti, E.Sacco, in

publication, *Materials Performance.*

4. S.Rossi, P.L.Bonora, L.Benedetti, M.Draghetti, C.Colombrita, E.Sacco, 'Proc. 0MC95', Ravenna Italy, March 1995, p.221.
5. S. Rossi, P.L. Bonora, R. Pasinetti, L. Benedetti, M. Draghetti, E. Sacco; submitted to publication, *Corrosion.*
6. W.H.Hartt, C.H.Culberson, S.W.Smith, *Corrosion Science,* 1984, **40,** 61 0.
7. S-H.Lin, S.C.Dexter, *Corrosion Science,* 1988, **44,** 615.

12. ELECTROCHEMICAL EVALUATION OF COATINGS FOR MARINE CORROSION CONTROL

P.L.Bonora, F.Deflorian, L.Fedrizzi, S.Rossi

Department of Materials Engineering, University of Trento
Via Mesiano 77, 38100 Trento, Italy

Abstract

The long term protection of metal structures exposed to a marine environment is a problem which needs to be approached from many different points of view involving several disciplines. A reliable construction incorporating a good metal substrate, suitable surface treatment, high performance coating, carefully designed cathodic protection, detailed maintenance plan and, last but not least, a complete testing and survey schedule can provide a safe and long service life. The use of electrochemical impedance spectroscopy (EIS) for the study of marine corrosion systems (composed of the metal, possible pretreatments, organic coating cycle, mechanical impacts and an aggressive environment) has allowed the relationship between some of the physicochemical parameters with the values of the corresponding components of the equivalent electrical circuit and the trend towards degradation to be determined. In particular, water absorption, porosity, delamination and the amount of undermining corrosion have been evaluated as a function of either natural or artificially induced degradation phenomena, including mechanical deformation. Another important field of application of E1S techniques regarding coatings and marine environment is the study of the features and performances of calcareous coatings obtained on an immersed structure as a consequence of cathodic protection.

1. INTRODUCTION

Electrochemical impedance spectroscopy (EIS) has been successfully used for the electrochemical characterisation of protective coating on metals [1-4]. As a first example of such coatings, we consider protective paints that are very often used as anti-corrosive media to protect marine structures. The system, which consists of a metal covered by an organic film, is generally quite complex and it may involve a large number of different situations. Firstly, the metal substrate may include different alloys: mild steel, aluminium, zinc (as a metal coating on steel sheets) or magnesium, etc.. These substrate surfaces can be pretreated in many different ways before painting e.g. chemical conversion treatments for coil coating (chromatisation, phosphatisation, new "chromium free" pretreatment, etc.) or mechanical surface preparations such as sand blasting or brushing. Finally the protective coating cycle often consists of paint layers which include adhesion promoters, surface *tolerant* primers, intermediate layers and top coatings. These organic layers have varying chemical and physical properties due to the different chemical compositions of the matrix and the presence of anticorrosive pigments.

All of these parameters (metal substrates, surface pretreatments, painting cycles) can influence the electrochemical behaviour as measured by EIS, which is also a function of the environment and the general conditions of measurement (temperature, oxygen concentration etc.)

The final objective of the EIS characterisation of protective organic coatings is to obtain information about the system properties such as presence of defects, reactivity of the interface, adhesion and barrier properties to water, etc. A knowledge of these parameters is extremely useful for the prediction of the in-service anti-corrosive behaviour of paints.

It is clear that to obtain these results it is necessary to identify the impedance values as a function of frequency as separate contributions due to the components of the system, for example discriminating between the contribution of the coating to the impedance from that of the substrate. In other words, equivalent electrical circuits which can model the impedance results need to be proposed. It is possible to choose a limited number of electrical elements (capacitance, resistance etc.) with an impedance equivalent to the impedance of the system studied for every frequency. A detailed discussion on equivalent electrical circuits can be found in specific works [5,6].

A similar approach can be used for studying calcareous coatings obtained on immersed structures under the effects of cathodic protection. A sacrificial anode corrosion protection system for marine structure is sized according to the protection current needed to maintain the polarisation of the entire structure at a level sufficient to render corrosion negligible. The current density needed for this is a function of the nature of the surface structure, the sea conditions, the electrochemical properties of the anodic alloy and the anode geometry.

The surface state of an immersed and cathodically protected structure is strongly influenced by the presence of any calcareous deposits which precipitate as a result of polarisation. The effect of such deposits is to limit current circulation in the galvanic element which the short circuited anodes and structure create.

A new sacrificial anode with a composite structure has been developed[7] for use in the protection of offshore structures against corrosion. In composite anodes, an external magnesium skin provides the high initial currents needed. After complete dissolution of the Mg, the remaining aluminium offers cathodic protection for the rest of the structure's lifetime, it being already polarised and covered with a good calcareous layer.

The initial polarisation produced by magnesium gives a more compact surface encrustation of the protected steel surfaces than that obtained with aluminium anodes. Thus, the use of a magnesium skin on aluminium anodes has the twofold effect of allowing rapid polarisation and of producing more compact deposits which limit current requirements together with a reduction in the mass of aluminium necessary[8,9].

2. MATERIALS AND EXPERIMENTAL PROCEDURES

The electrochemical measurements were performed on different organic coatings. The samples were cold rolled low carbon steel sheets or galvanised steel sheets with different industrially poduced organic coatings,

- fluoropolymer coatings
- epoxy coatings
- polyester coatings

with thicknesses ranging from about 30 to 100 μm.

Electrochemical measurements were also made on samples with a defect caused by puncturing with a loaded stylus. The reproducibility and the dimension of the defect was checked by observations through the light microscope. Two values for the diameters values of the induced defect were obtained, *ca* 100 μm and 300 μm.

The electrochemical impedance measurements were performed at the corrosion potential in the 10 kHz - 1 mHz frequency range using a Solartron 1255 Frequency Response Analyser connected to a 273 PAR potentiostat.

The amplitude of the sinusoidal voltage signal was maintained at 50 mV for undamaged samples until a corrosion attack appeared and then it was changed to 10 mV; for the samples with an artificial defect, it was 10 mV from the beginning. The electrochemical tests were carried out in quiescent aerated 3.5% sodium chloride aqueous solutions, simulating the marine aggressive environment, or 0.3% Na_2S0_4 for studying the coatings with artificial defects. The exposed painted sheet area was about 30 cm^2.

The three electrode electrochemical cell was obtained by the adherion of a plastic cylinder on the sample sheet and filling it with the test solution. A saturated sulphate reference electrode (SSE = + 640 mV vs.SHE) and a platinum counter electrode were employed.

For studying calcareous coatings, a tank with 50 dm^3 artificial sea-water (ASTM D 1141-90) was used. Laminar flow over the immersed surfaces at speeds between 3 and 10 $cm\,s^{-1}$ was maintained by means of a thermocryostat (flow 20 $dm^3\,min^{-1}$). The temperature was 14 °C (the average temperature in the North Adriatic Sea). The pH 7.9 was maintained during the tests by the addition of Hcl. The bicarbonate content was held constant by daily additions of sodium bicarbonate. One of the plate faces was completely insulated with silicone, whilst the exposed surface was cleaned with 320 emery paper and acetone.

For a first period of 18 days, the steel plates (Sample 1) were polarised at -870 mV vs. Ag/AgCl (protection condition with Al anode) and others (Sample 2) at -1120 mV vs. Ag/AgCl (Mg anode protection condition). Over a second period of 10 days, all samples were held at -870 mV. Measurements of the electrochemically polarised impedance were made by superimposing a 10 mV sinusoidal signal at frequencies from 100 kHz to 1.5 mHz on the cathodic potentials selected (-870 and - 1120 mV).

These laboratory tests included the analysis of the protective capacity and of the structure of the calcareous deposits formed on the cathodes *via* macro- and microscopic observation, electron microscopy and X-ray diffraction using an X Rigaku diffractometer Model RAD B 3.

3. RESULTS AND DISCUSSION

3.1 Organic Coating Defect Area Determination

Awareness of the area of defects existing in an organic coating, through which a corrosive process can occur, is one of the more important pieces of information needed in order to understand the protective properties of the coating[10,11]. The interaction between the metal and the aggressive electrolyte is due to the presence of defects or to the intrinsic porosity. This is mainly true in the case of coatings having high barrier properties. The electrochemical impedance results can be studied by using an equivalent electrical circuit modelling the clectrochemical behaviour of the system. A very popular equivalent electrical circuit used for organic coated metals is shown in Figure 1. In this, the coating capacitance Cc, the coating resistance Rp (related to the ionic barrier properties of the coating) the double layer capacitance Cdl, the charge transfer resistance Rct (related to the corrosion rate) and the electrolyte resistance Ro are represented. A method for the evaluation of the defect area on the basis of the analysis of the electrochemical Impedance diagram was proposed by Haruyama[12].

The theoretical bases of this method, also known as the break-point method, are well described elsewhere[13,14]. The diagram evaluation leads, as a final result, to a direct proportionality between the ratio Ad/A (where A is the total area tested and Ad the area of the defects) and f_{45}, see equation (1),

$$f_{45} \rightarrow K\frac{Ad}{A} \qquad (1)$$

where,

$$K \rightarrow \frac{1}{(2\pi\varepsilon\varepsilon_o\rho_o)} \qquad (2)$$

and ε represents the relative permittivity of the layer, ε_o that of free space and ρ_o the specific restivity. This is the point, in the high frequency range, at which the phase of the impedance vector reaches 45 degrees The accuracy in calculating the f_{45} value is mainly a function of the ratio of the capacitances and the accuracy particularly increases if the

Cdl/Cc ratio, increases, with the only hypotheses being that Ro is negligible and Rct ≥ Rp, which are satisfied in the usual systems studied. In the initial stage of degradation, when the defect area is very small, the values of Cc and Cdl can be similar and therefore it is impossible in this case to correctly measure the pore area from the high frequency f_{45} break-point[15,16].

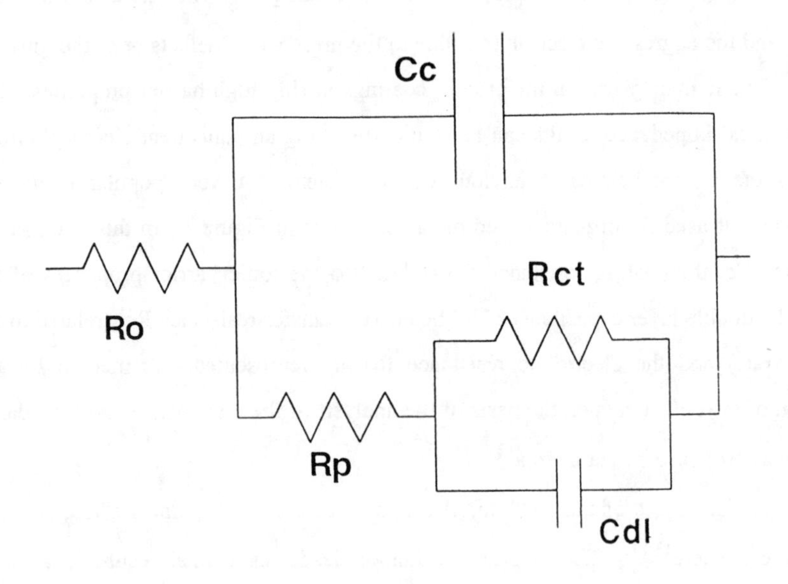

Figure 1 *Equivalent electrical circuit for a metal with an organic coating.*

Another method of determining the area of the defects is to evaluate the pore resistance (Rp) which can be obtained by interpolating the experimental data using suitable calculation programs[17].

$$R_p \to \frac{\rho_0 d}{Ad} \tag{3}$$

Hence, knowing d and ρ_o it is possible to determine the Ad value. The main problem in the determination of Ad, following both the above mentioned methods, is the evaluation of ρ_o.

Another alternative method of determining the area of the coating defects is the analysis of the low frequency region of the impedance diagram, where a second time constant is clearly visible. This is representative of the faradaic reaction and allows the calculation of Rct and Cdl for the coated samples. Knowing the Rct' or Cdl' values for a unit of bare metallic area, which can be obtained by impedance measurements on an uncoated metal sheet, it is possible, by a simple ratio, to obtain an approximate value of the metal area subjected to the corrosive reactions.

$$\textit{Reactive area} \rightarrow \frac{Rct'}{Rct} \rightarrow \frac{Cdl}{Cdl'} \qquad (4)$$

Figure 2 shows the impedance spectrum for a fluoropolymer coated steel sheet containing an artificial defect, diameter of about 100 μm, obtained after 1 hour of immersion. The dimension of the artificial defect is wide compared to the thickness of the organic coating (about 30 μm), hence, the value of ρ_o has been assumed to be the same as the specific resistivity of the bulk solution. This value, obtained by conductivity measurements carried out at 25 °C, is about 300 Ω cm^{-1}. Therefore it is possible to calculate the diameter value of the defect by equation (3). The value of 77 μm obtained is in good agreement with the real value of 100 μm.

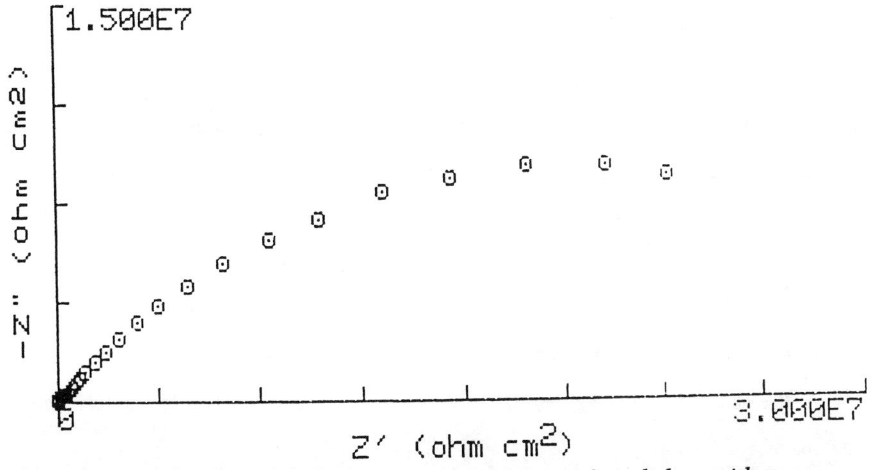

Figure 2 *Impedance diagram of a fluoropolymer coated steel sheet with an artificial defect with a diameter of about 100 μm obtained after 1 hour of immersion.*

By assuming the same value of ρ_o, it is also possible to calculate the defect area using impedance data for samples with 300 μm diameter artificial defects. In this case, after 4 hours of immersion, the value is about 211 μm, close enough to the real value of 300 μm.

These preliminary tests confirmed the reliability of electrochemical impedance results in determining the dimensions of defects in organic coatings with a reasonable approximation. Only one hypothesis was made to calculate the defect area i.e. that the value of ρ_o is the same as the specific conductivity of the bulk electrolyte. This hypothesis is certainly true in the case of a defect of wide dimensions, in the range of those used in our study. The evaluation of ρ_o becomes more difficult for materials having a smaller defect diameter; in this case ρ_o may be different from the value of the conductivity of the bulk electrolyte.

However in our study, in agreement with other authors[18], we assumed, as a first approximation, that the value of ρ_o in equation (3) is constant for all the samples analysed, including those without artificial defects. The correct evaluation of ρ_o remains, without doubt, a critical step in the use of equation (3) and this aspect requires a thorough experimental examination in the future. The defect measurements, for instance, show the lower porosity in the fluoropolymer sample (about 0.2 μm^2 cm^2) in comparison to similar samples without fluorine (about 24 μm^2 cm^2). This data confirms the better protective properties of the fluoropolymer coatings.

The use of a fitting procedure also gives us the opportunity to measure the values of Rct. From a knowledge of the value of Rct for the bare metal, which in the present case is about 1250 Ω cm^2, it is possible to calculate the substrate active area from the ratio between the value of Rct per unit area and that of Rct obtained by the fitting of the impedance experimental data on the organic coating. It is interesting to note in Figure 4 (fluoropolymer coating with 100 μm diameter region of artificial defect) that the values of the active area, measured by the evaluation of Rct, increase with the time of immersion, whereas in the same time the values of the defect area, calculated by Rp values, remain almost constant.

The two values coincide at zero time, then move away from each other. This trend is due to the widening of the reactive area all around the artificial defect, as a consequence of the disbonding of the organic coating (Figure 4). This hypothesis was confirmed by microscopical observation of the samples after testing.

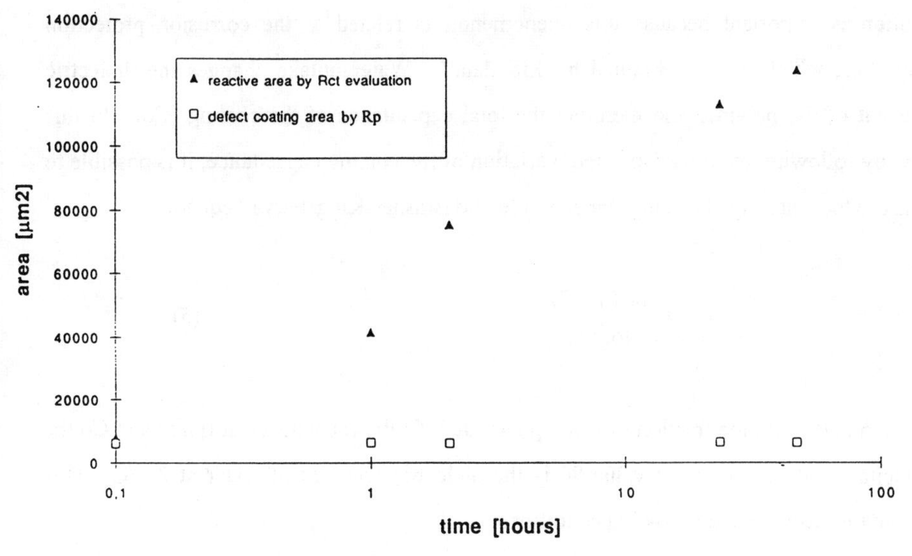

Figure 3 *Defects area and reactive area for an organic coating with artificial defect.*

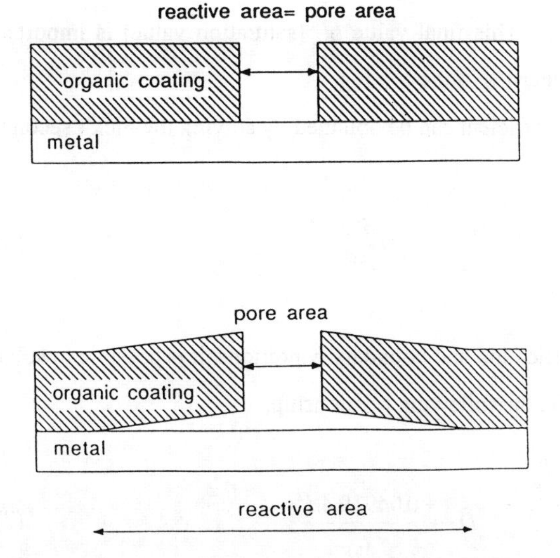

Figure 4 *Delamination process around a defect in an organic coating.*

3.2 Organic Coating Water Uptake

The evaluation of the water uptake process into an organic coating immersed in a solution is important because this phenomenon is related to the corrosion protection properties, which can be obtained by EIS data[19]. Water uptake changes the dielectric constant of the polymer and therefore the total capacitance of the coating (Cc). In this way, by following the time dependent variation of the coating capacitance, it is possible to measure the water uptake using, for example, the Brasher-Kingsburys[20] equation,

$$\Phi \rightarrow \frac{\log Ct / Co}{\log 80} \tag{5}$$

where ϕ is the volume fraction of water penetrated, Ct the capacitance at time t and Co the capacitance at time t=0; the value 80 is the dielectric constant of water at 25 °C. This formula is based on the following conditions:

a) the water is homogeneously dispersed;

b) the absence of water-polymer chemical interactions;

c) penetration of a low volume and no swelling of the matrix[19].

The typical evolution of ϕ *vs.* time shows an increase after immersion, reaching a constant value after some hours. This final value ϕ_s (saturation value) is important in order to evaluate the barrier properties of the coating.

The diffusion coefficient can be obtained by solving the Fick's second law,

$$\frac{\delta c}{\delta t} \rightarrow -D\frac{\delta^2 c}{\delta x^2} \tag{6}$$

Assuming some simplification and boundary conditions, it is possible to solve equation (6) and obtain, for example, the following relationship,

$$D \rightarrow \frac{0.04919.4d^2}{t_{0.5}} \tag{7}$$

where $t_{0.5}$ is the time needed to reach a value of ϕ which is half of the saturation value ϕ_s

and d is the coating thickness. From a ϕ *vs.* time plot it easy to find this parameters and therefore to calculate D. Many other mathematical methods, making use of equations different from equations (6) and (7), are available in the literature for determining both water content and the diffusion process[19].

Many authors[21,22] have studied the water uptake process by impedance measurements, with the evaluation of the trend of the coating capacitance. For epoxy coatings, this increases after immersion in an electrolyte, reaching a constant value after some hours[19]. A subsequent increase of the capacitance value after a long lasting plateau period shows the beginning of detachment of the organic coating from the substrate due to adhesion loss[21].

Prior to saturation, the trend of the water uptake follows Fick's first law. Plotting the water content obtained using the Brasher and Kingsbury equation, as a function of the square root of the time, shows a linear relationship before the beginning of the saturation process. In Figure 5 this trend is shown for a polyester coating 30 μm thick in 3.5% NaCl solution. From these data it is possible to obtain the value of water uptake saturation (about 4.5 %) and the diffusion coefficient, using equation (6). In our example the diffusion coefficient is about $6.34 \times 10^{-13}\ m^2\ s^{-1}$.

Another interesting example of the water uptake determination is the comparison of epoxy coated galvanised steel, with and without artificial defects. The initial values of capacitance measured for the undamaged samples are about 1.3×10^{-10} F cm^{-2} for the thicker coating (100 μm) and about 3.5×10^{-10} F cm^{-2} for the thinner one (30 μm). Figure 6 shows the trend of Cc as a function of time. Both samples indicate the presence of a plateau and a subsequent increase of the capacitance value in agreement with the mechanism of water uptake and detachment of the coating[19]. It is important to observe that the beginning of the coating detachment occurs at different times as a function of the coating thickness (about 50 days for the thinner coating and 75 for the thicker one). Using the Brasher and Kingsbury formula[20], water absorption in the 30 μm coated sample after 15 days of immersion is 2.11%. In contrast, the samples with an artificial defect (30 μm thickness) show a continuous increase of the capacitance value after immersion (Figure 7).

The evaluation of the water content due to water absorption in the coating after 15 days of immersion indicates a value of about 5.56%. Such a high value[21], with the anomalous capacitance trend (continuous increase with immersion time), can be justified by assuming water penetration (delamination) which can occur between the coating and the substrate in *correspondence* with the perimeter of the artificially produced defect in the coating.

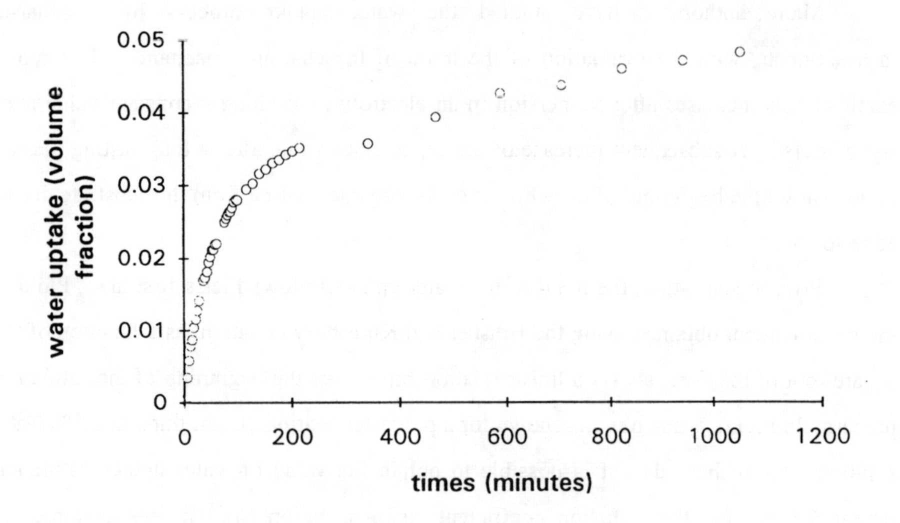

Figure 5 *Water uptake for a polyester coating of 30 μm of thickness.*

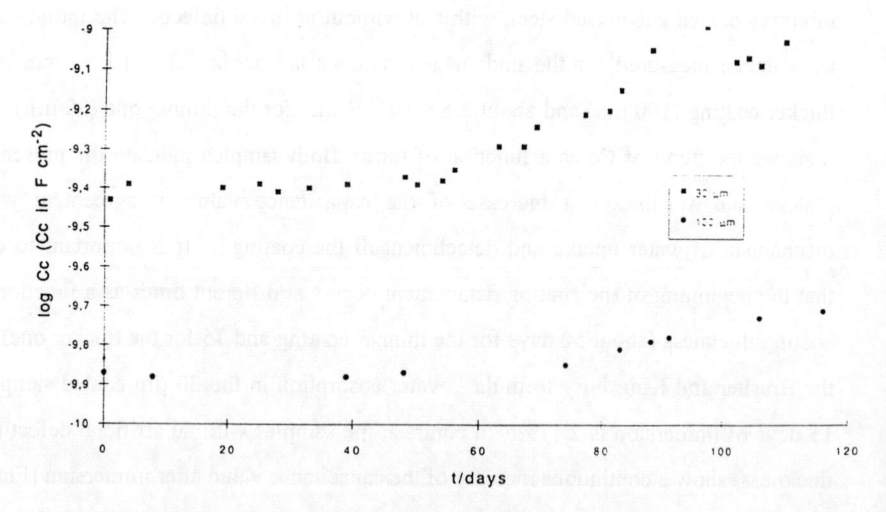

Figure 6 *Coating capacitance trend for an epoxy coating.*

3.3 Evaluation of Calcareous Coatings

Impedance measurements were carried out on polarised steel samples at typical values (-870 mV and -1120 mV) which describe the various reactions that sustain cathodic depolarisation on steel plates[23].

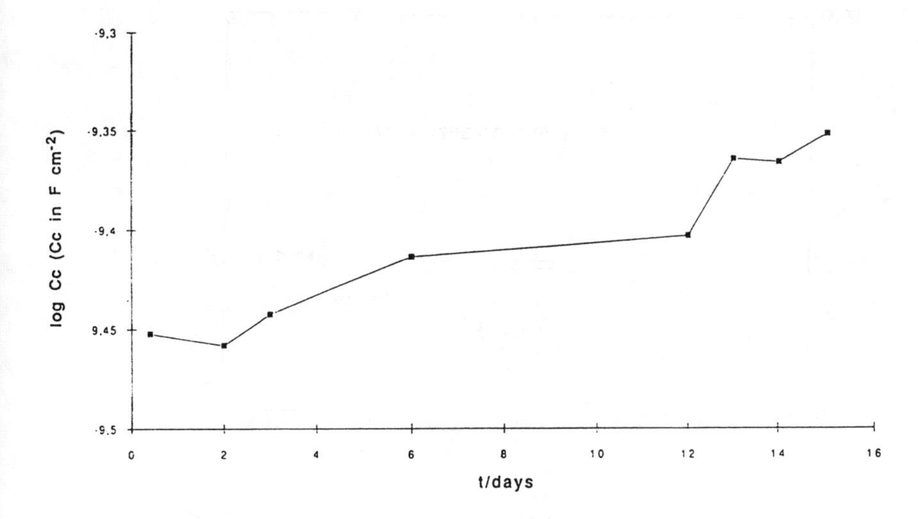

Figure 7 *Coating capacitance trend for an epoxy coating with artificial defect.*

Figures 8 and 9 show the impedance *vs.* time diagrams, as Nyquist plots: the charge transfer resistances for these two reactions differ considerably, the electrochemical reaction on Sample 1 at -870 mV controlled by oxygen diffusion in the electrolyte, whilst for Sample 2 at lower polarisation there is a cathodic water reduction reaction with activation control. Both samples show, in time, an increase in resistance during the reduction reaction, due to a decrease in the reactive areas and to the greater protective capacity of the deposits formed. This variation, on the steel Sample 1 at -870 mV, leads to a rapid drop in current supplied even over the first two days of the experiment, whilst on Sample 2 at -1120 mV, similar current values are not reached (Figure 10). Such current densities, after the initial polarisation peak and initial growth of deposits, remain between 1200 and 3000 mA m^2, compared with values of 100-150 mA m^2 with steel polarised at -870 mV. This condition confirms that Mg dissolves very rapidly when used for cathodic protection in sea-water; as well as the low anodic efficiency, the cathodic potential developed produces high currents; at this potential, the cathodic reaction is water reduction which is not very hampered by the presence of calcareous deposit. The action of Mg does however produce a deposit which is an effective barrier to oxygen diffusion when the potential is -870 mV where the cathodic reaction is oxygen reduction.

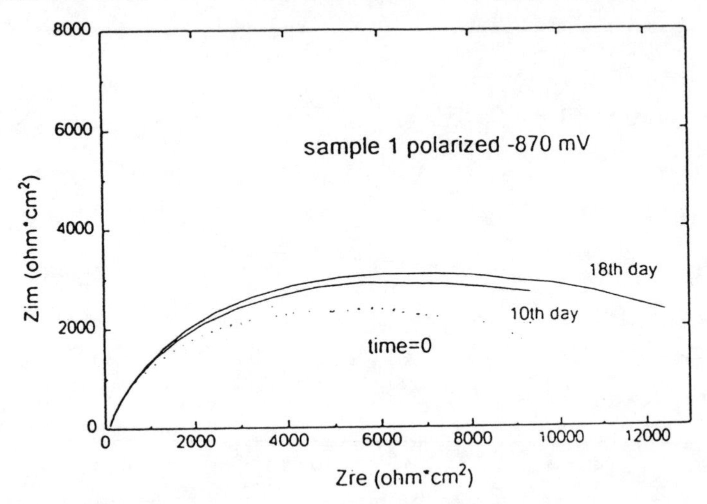

Figure 8 *Impedance measurements on polarised steel samples at -870 mV (Ag/AgCl)*

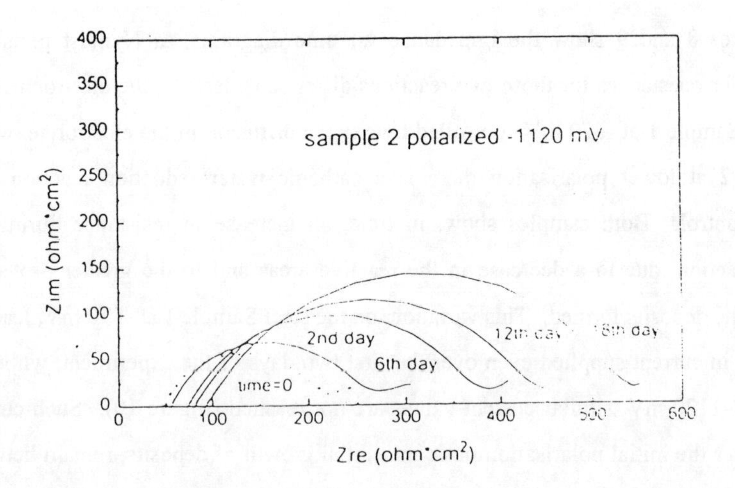

Figure 9 *Impedance measurements on polarised steel samples at -1120 mV (Ag/AgCl)*

This effect is clearly visible in the diagram in Figure 10, where we can see that after increasing the potential of Sample 2 to -870 mV from - 1120 mV, the associated current rapidly reaches values between 30 and 50% (25-35 mA m^2) of the values of the current for Sample 1 (65-85 mA m^2). This is confirmed by the impedance measured after increasing the potential: on Plate 2 the charge transfer resistance becomes much higher than that of Sample 1 which has now stabilised after being polarised from the start at -870 mV (Figure 11).

During the ten days of the second phase of this experiment, from the moment the Sample 2 potential was increased, no results were obtained from the measurements of current and impedance.

The SEM and XRD data obtained show the presence of a compact orderly needle-like aragonite structure in globule form, for the deposit formed on steel initially polarised at - 1120 mV (typical potential of Mg anode) while the deposit formed during the test at -870 mV (potential of Al anode) is more disordered and less compact. The more intense peaks in X-ray analysis refer to aragonite and the presence of magnesium hydroxide was determined only in samples initially polarised at-1120 mV.

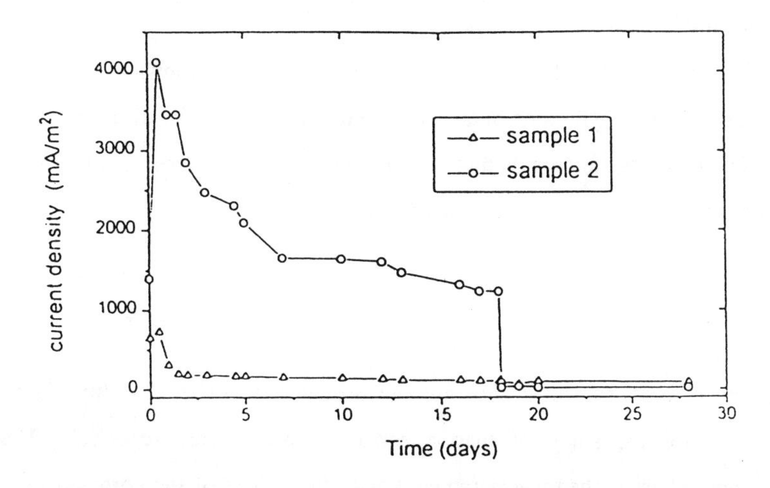

Figure 10 *Current density on steel samples*

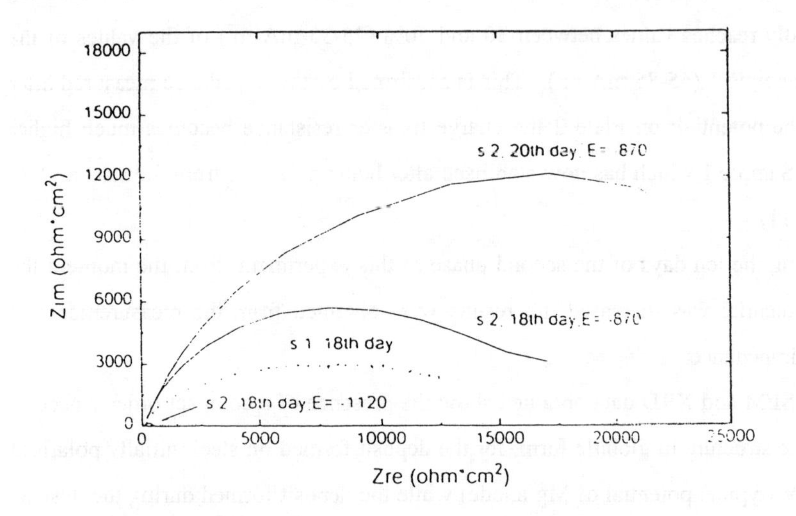

Figure 11 *Impedance plot of samples 1 and 2 after 18 days of immersion*

The analyses made allow us to reach the following conclusions:

- the deposit formed under the potential of the action of Al anodes includes calcium carbonate showing low barrier properties
- the deposit formed under the adequate action of composite anodes shows the presence of calcite and magnesium hydroxide in traces amounts when/if present, while aragonite which characterises the microstructure of the deposit guarantees good barrier characteristics and stability of the encrustation with time.

4. CONCLUSIONS

In this paper some examples showing the use of Electrochemical Impedance Spectroscopy (EIS) for the study of marine corrosion systems are reported. The relationship of some physico-chemical parameters with the values of the corresponding components of the equivalent electrical circuit and with the trend towards degradation is considered. In particular, water absorption, porosity, delamination and the amount of undermining corrosion for organic coated metals were evaluated as a function of either natural or artificially induced degradation phenomena. Another important example of the application of EIS techniques to coatings for the marine environment is the study of the features and performances of calcareous coatings obtained on an immersed structure under the effects of cathodic protection.

References

1. F. Mansfeld, M. Kendig and S. Tsai, *Corrosion,* 1982, **38**, 478.
2. F. Deflorian, L.Fedrizzi and P.L. Bonora, *Prog. Org. Coat.,* 1993, **23**, 73.
3. M. Kendig and J.Scully, *Corrosion,* 1990, **46,** 22.
4. J. Titz, G.H. Wagner, H.Spaehn, M.Ebert, K. Juettner and W.J. Lorenz, *Corrosion,* 1990, **46,** 221.
5. F. Deflorian, L.Fedrizzi and P.L. Bonora, *Electrochimica Acta,* 1996, in press.
6. E.P.M. van Westing, G.M. Ferrari and J.H.W. de Wit, *Corrosion Sci.,* 1994, **36,** 1323.
7. S. Rossi, P.L. Bonora, R. Pasinetti, L. Benedetti, M. Draghetti and E. Sacco, *Materials Performance,* 1996, in press.
8. P.L. Bonora, M. Draghetti, C. Colombrita, A. Ghisetti and G.L. Valla, 'Proceedings UK Corrosion', London, 1993, Vol 1.
9. S. Rossi, P.L. Bonora, L. Benedetti, M. Draghetti, C. Colombrita and E. Sacco, 'Proceedings OMC 95', Ravenna, 1995, p. 221.
10. F.M. Geenen, J.H.W. de Wit, E.P.M. van Westig, *Prog.Org.Coat.,* 1990, **18,** 299.
11. A.Amirudin, D.Thierry, *Br.Corros.J.*, 1991, **26**,195.
12. S. Haruyama, M.Asari and T.Tsuru in 'Corrosion Protection by Organic Coatings', M.Kendig and H.Leidheiser eds. Proc. Vol. 87-2, Pennington, NJ: Electrochemical Society [ECS], 1987, p. 197.
13. J.R. Scully, *J.Electrochem.Soc.,* 1989, **136,** 979.
14. R. Hirayama, S. Haruyama, *Corrosion,* 1991, **47,** 952.
15. F. Deflorian, L.Fedrizzi, P.L. Bonora, *Electrochim. Acta,* 1993, **38** 1609.
16. F.Deflorian, L.Fedrizzi, P.L.Bonora, *Corrosion,* 1994, **50,** 113.

17. B. Boukamp, *Solid State Ionics,* 1986, **20,** 31.
18. R.D.Armstrong and D.Wright, *Electrochim. Acta,* 1993, **38,**1799.
19. F.M. Geenen, PhD Thesis, University of Delft, 1991.
20. D.M. Brasher and A.H. Kingsbury, *J. Appl. Chem.* 1954, **4,** 62.
21. F.Deflorian, V.B.Miskovic-Stankovic, P.L.Bonora and L.Fedrizzi, *Corrosion, 1994,* **50,** 446.
22. T.Monetta, F. Bellucci, L.Nicodemo and L.Nicolais, *Prog. Org. Coat.,* 1993, **21,** 353.
23. B.J.Little and P.A. Wagner, *Materials Performance,* 1993, **9,** 16.

13. METAL FILLED COMPOSITES AS PROTECTIVE COATINGS AGAINST MARINE CORROSION

N. Kouloumbi, G.M. Tsangaris, S.Kyvelidis

National Technical University, Chemical Engineering Department.
Materials Science and Engineering Section, 9, Iroon Polytechniou Str,
Athens 157 80, Greece

Abstract

The effect of the presence of metal powders in an epoxy coating and the influence of their concentration on the assessment of the coatings behaviour in marine environment has been studied. Pretreated steel specimens have been coated with a layer of metal filled epoxy resin and their electrochemical and dielectric behaviour in corrosive environment (3.5% wt NaCl) has been investigated. Electrochemical Impedance Spectroscopy and Dielectric measurements were performed. In an effort to examine the influence of the presence of more and less electroactive metal powders than the steel substrate, Zn, Al, Fe, Cu, Ni in powder form were used as fillers in the epoxy coatings. Metal powder addition does not generally seem to worsen the protective performance of the coatings, compared to coatings of pure epoxy resin, except for the composites with copper or nickel particles which show a reduced protection performance, but they are still protective.

1. INTRODUCTION

Epoxy resins used as binder for industrial coatings have become increasingly important representing perhaps the best combination of corrosion resistance and mechanical properties, thus providing outstanding service under severe conditions[1-3]. Recent increasing technical demands justify recognition of composite coatings, such as coatings reinforced with glass, polyaramide or graphite fibres which exhibit improved thermal and mechanical properties[4,5]. Additives incorporated within the coating offer protection either by exerting a barrier effect or acting as a sacrificial anode or as an inhibition agent. Conductive fillers can greatly alter conductivity and dielectric characteristics of the polymeric matrix in which they are dispersed, affecting significantly the diffusion of the water to the metal-coating interface and consequently the appearance of galvanic interactions. Zinc rich coatings are commonly used and offer protection[6], however the study of other metallic fillers in polymeric matrices presents a challenge since either the metallic powders or their corrosion products may alter the protective performance of the coating.

In the present work composite particulate coatings of epoxy resin and metal powders of Zn, Al, Fe, Cu and Ni were prepared and used for the protection of steel surfaces against marine corrosion. The role of the powdered metals on the performance of the coatings was investigated. The transport pattern which determines the conductivity in such conductive polymeric composites undergoes an insulator conductor transition which depends on the concentration of the particles. For this reason experiments were performed for a concentration of 15% w/w common to all metal powders but for the iron powder two more concentrations namely 7.5% and 30% w/w were studied.

2. EXPERIMENTAL DETAILS

The test specimens were circular disks (diameter ϕ=6cm) of low carbon steel. The coating used was a commercially available bispenol-A-type epoxy resin (DER 321, DOW Chemical Co.) cured with a cycloaliphatic amine (product YZ 87706.65 Dow Chemical Co.).

The above system was used wither in this formulation or filled with the powdered metals in a content of 15% w/w and 7.5%, 15%, 30% w/w for iron. Powdered metals were incorporated in the liquid state at 30 °C. The dry film thickness at each surface was equal either to 70±5 or 250±20 μm. Coated and uncoated specimens were immersed in 3.5% NaCl solution for predetermined immersion times. Details concerning preparation of specimens instrumentation and measurements can be found elsewhere[7,8].

3. RESULTS AND DISCUSSION

The corrosion tendency of the steel substrate was estimated in all types of coated specimens, by the half-cell potential development versus the exposure time to the corrosive environment. In all cases, as shown in Figures 1 and 2, specimens coated with composites of epoxy resin and aluminium, iron or zinc powder, immediately after immersion in solution, exhibit a potential value which corresponds closely to that of the specimens coated with pure epoxy. During the first days of immersion a decay to more negative potential values was observed followed by a tendency to stabilise to a steady state value around –600 mV. The trend to reach stable values that differ less than 100 mV from that of the free corrosion potential developed by uncoated steel, implies an increasing electrochemical activity of the metal surface during the test. However, the similar free corrosion potential evolution and the small differences in the corrosion potential plateau values irrespective of metal type or iron powder concentration do not enable any qualitative prediction of a different corrosion performance of any type of coated specimens tested.

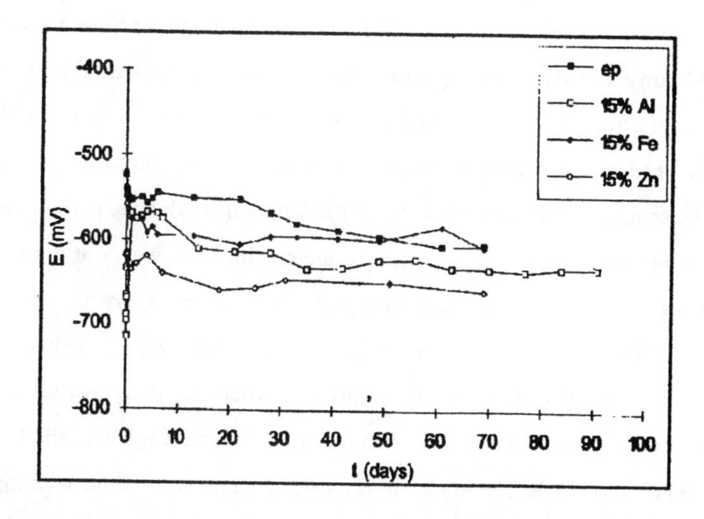

Figure 1 *Half-cell potential vs immersion time for different types of epoxy coated specimens*

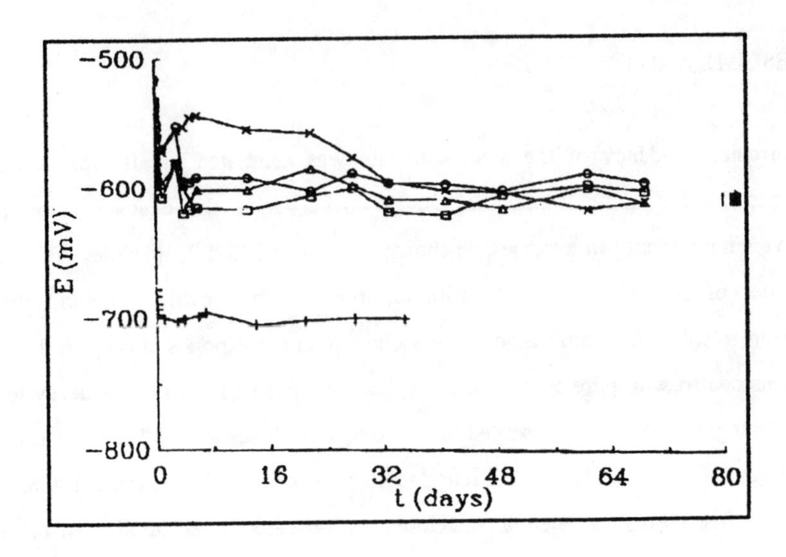

Figure 2 *Time dependence of corrosion potential for different types of coated and uncoated steel specimens: +, bare steel; x, pure epoxy; Δ, 7.5%; O, 15%; □, 30 % wt/wt. iron powder.*

The anticorrosive performance of the coatings was evaluated by potentiodynamic polarisation measurements (Figures 3, 4). At the beginning of the exposure the addition of Fe, Zn and Al powder leads to coating with very low corrosion rates similar or somewhat lower than that of a pure epoxy resin coating. The dispersion of conductive particles in an epoxy matrix changes the transport properties of the coating because it leads to the decrease of the coating resistance and to the easy creation of conductive paths affecting in this way negatively the protective properties of the coating[9]. However at the same time, the metal powder grains, by filling the micropores of the polymeric matrix, exert a barrier effect which positively affects the protective properties of the coating[10] and consequently specimens with coatings containing these metal powders or iron powder in various concentrations, show improved corrosion behaviour. In the coatings with the lowest iron powder content (5 % w/w) the improvement is clearer than in the coatings with higher iron powder content (15 and 30% w/w) whose corrosion performance is close to that of the pure epoxy resin. This behaviour can be thought of, as the result of a counterbalance between the barrier effect exerted by the composite coatings and a possible unsatisfactory wetting of the metal particles by the resin and/or a lower resistance of resin/metal particles adhesion to wet condition, as the metal powder concentration is increased. After 90 days of immersion to the corrosive environment the corrosion performance of specimens with coatings containing metal powder remains close to that of specimens coated with pure epoxy resin (Figures 5 and 6).

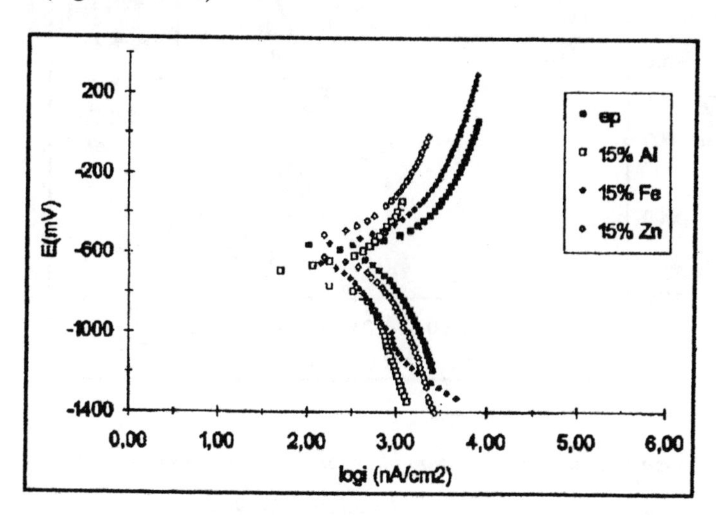

Figure 3 *Potentiodynamic polarisation plots for different types of epoxy coated steel Specimens just after the immersion*

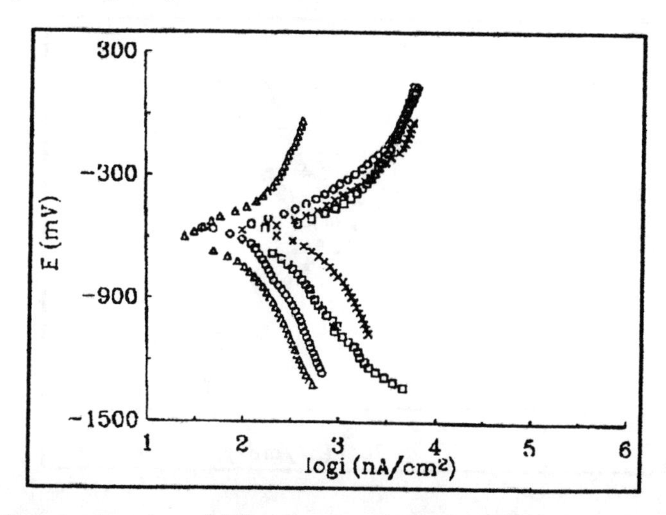

Figure 4 *Polarisation plots of different types of coated steel specimens exposed to 3.5% NaCl for 4h: x, pure epoxy; Δ, 7.5%; O, 15%; □, 30% w/w iron powder.*

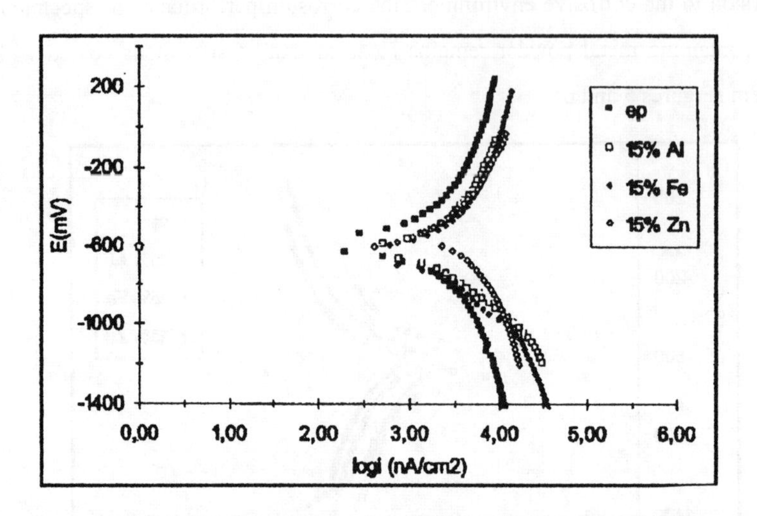

Figure 5 *Potentiodynamic polarisation plots for different types of epoxy coated steel specimens exposed to 3.5% NaCl solution for 90 days.*

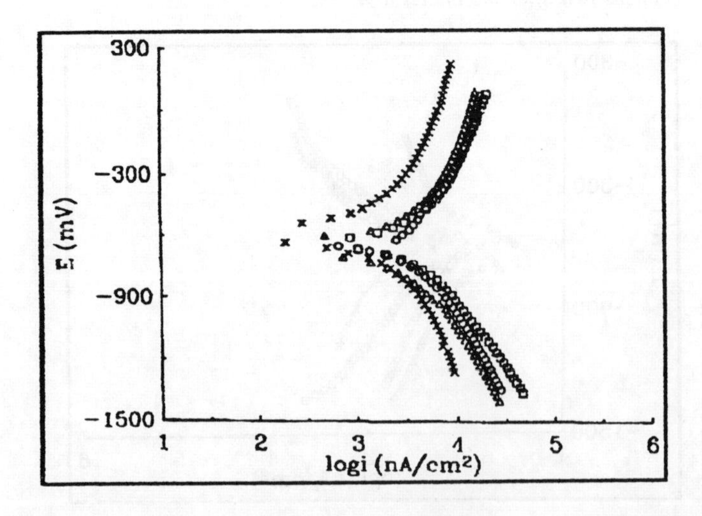

Figure 6 *Polarisation plots of different types of coated steel specimens exposed to 3.5% NaCl(90 days): x, pure epoxy; Δ, 7.5%; O, 15%; □, 30% w/w Fe powder.*

Visual examination of the specimens reveals in the early stage of exposure, the appearance of green coloured spots which develop to greater black regions with the increase of the exposure time. The initially appeared spots are attributed to the precipitation of green complexes formed in a chloride environment as intermediate in the formation process of oxyhydroxides or iron oxides. There is a great possibility that the iron oxides compose the black products possessing a high adherence to the metal surface. These corrosion bulky products can create a barrier effect and can counterbalance the increase in permeability due to the presence or iron particles.

Consequently coatings of 70 μm thickness with Fe, Al and Zn exhibit almost similar protective behaviour which is not inferior to that of pure epoxy resin coatings for the immersion times of this work, although there is a small increase in the permittivity due to the presence of the metallic particles. Also the minor differences observed strengthen the assumption that the iron concentration levels used in these coatings are lower than the percolation threshold and thus the systems have not yet become purely conductive.

The addition of Ni, Cu and Fe in coatings of dry film thickness equal to 250 μm leads to protective coatings (Figure 7). Coatings with iron show the best results while coatings with Cu and Ni powders exhibit almost similar protective behaviour which is however inferior to that of unfilled epoxy coatings.

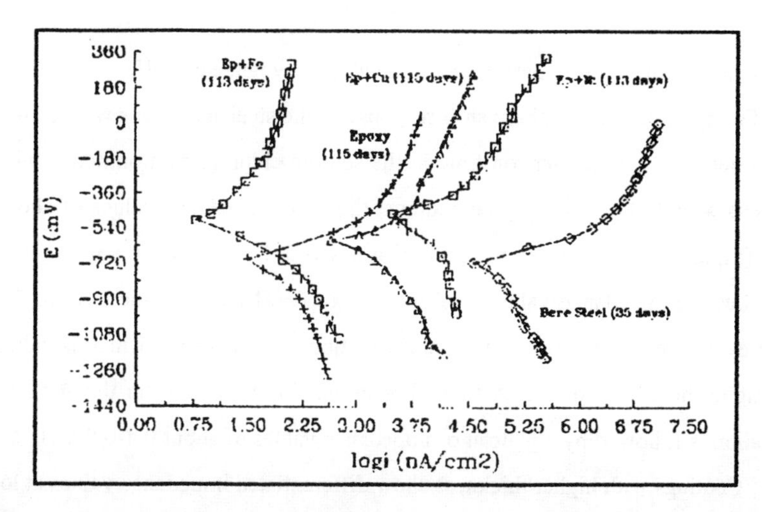

Figure 7 *Potentiodynamic polarisation plots for uncoated and epoxy coated steel specimens in 3.5% NaCl*

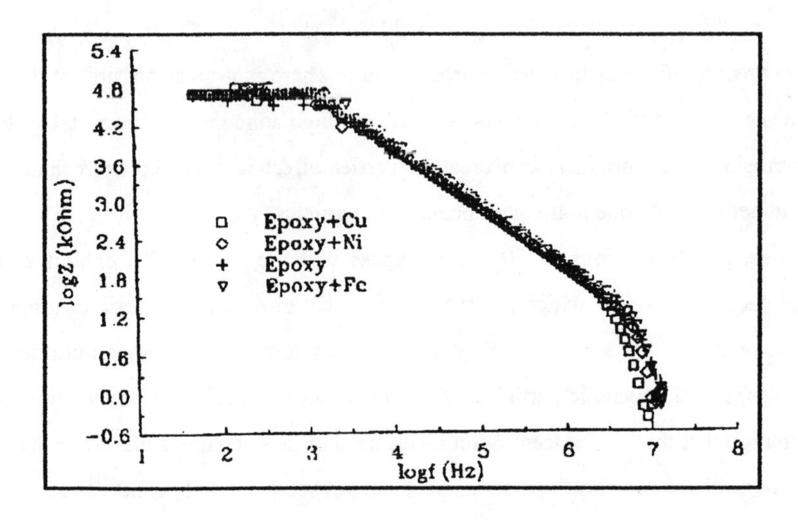

Figure 8 *Bode plots for uncoated and different types of epoxy coated steel specimens just after the immersion in 3.5% NaCl solution.*

EIS measurements were performed in the frequency range of 20Hz to 13 MHz. The modulus of impedance and the phase shift response of the steel specimens with composite coatings, is depending on immersion time in the sodium chloride environment. Initially, all specimens with 250 μm coating have almost the same impedance values of about 2 x 10^9 Ω cm^2 (Figure 8). This is also valid for coatings with Fe, Al and Zn with a thickness of 70 μm, but the impedance value is of about 2 x 106 Ω cm^2. In the case of 70 μm coatings at early immersion times, there is a fluctuation of these values which is indicative of the changing behaviour of the coating. For immersion times greater that 10-20 days this fluctuation is followed by a plateau of impedance values of about 0.4-0.8 x 10^9 Ω cm^2 (Figure 9). Coatings with higher thickness show differentiated impedance values at longer exposure times depending on the coating type (Figure 10). The coating with iron powder shows the least decrease in impedance, while those with nickel powder show the most.

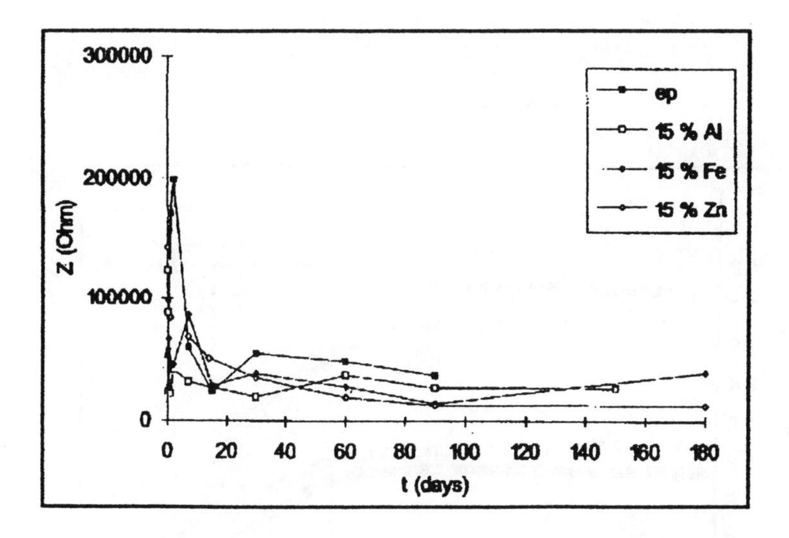

Figure 9 *Impedance vs immersion time for different types of epoxy coated steel specimens.*

As has been already mentioned, the addition of powders in the polymer affects its protective properties either by filling the micropores of the polymer and thus exerting a barrier effect which reduces the substrate corrosion rate, or by creating conductive paths, which in the case of nickel or copper addition, accelerates the steel substrate corrosion rate due to galvanic corrosion. In the case of iron powder, the absence of galvanic corrosion reveals the positive influence of micropore filling, either by the powder grains or by the bulky corrosion products. This assumption explains the behaviour shown, that is, the iron-filled coatings are the most protective, while those with nickel or copper, though still protective, exhibit lower protection than the epoxy without filler. This behaviour is strengthened by the fact that iron shows good catalytic properties in the cathodic oxygen reduction reaction,

$$\frac{1}{2} O_2 + H_2O + 2e^- \longrightarrow 2OH^- \qquad (1)$$

leading to accelerated oxidation of the metal powder and thus producing bulky corrosive products, which fill the micropores of the polymer and increase the coating resistance.

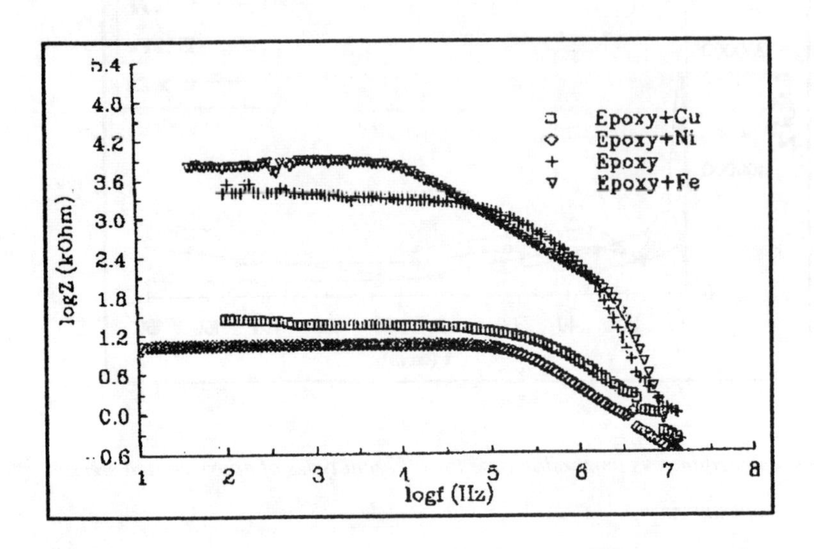

Figure 10 *Bode plots for uncoated and different types of epoxy coated steel specimens exposed to 3.5% NaCl solution for 115 days.*

At early exposure times two capacitive responses and consequently two maximum phase shifts are displayed for composite coatings of 70 μm thickness. This implies that in this case two time constants exist[11], while coatings of pure epoxy resin show one time constant (Figure 11).

The best fit of the EIS data and the simulated spectra are shown by the full lines. The equivalent circuit used is the common one considered by many authors for a non-barrier type coating which can express both the capacitive and resistance characteristics of the coating and the corrosion process of the iron filler of the composite. In the region of medium frequencies ($10^{2.5}$-$10^{4.5}$ Hz) (Figure 11), the time constant is indicative of the oxidation of the metallic grains used as filler, while in the high frequencies (>10^5 Hz) the time constant is attributed to the capacitive behaviour of the polymeric film. The absence of the medium frequencies time constant from the responses at long immersion times (Figure 12), should be associated with the evolution of corrosion products throughout the film which modify the characteristics of the coating. In this case the coating behaviour is expressed by the equivalent circuit for a barrier coating.

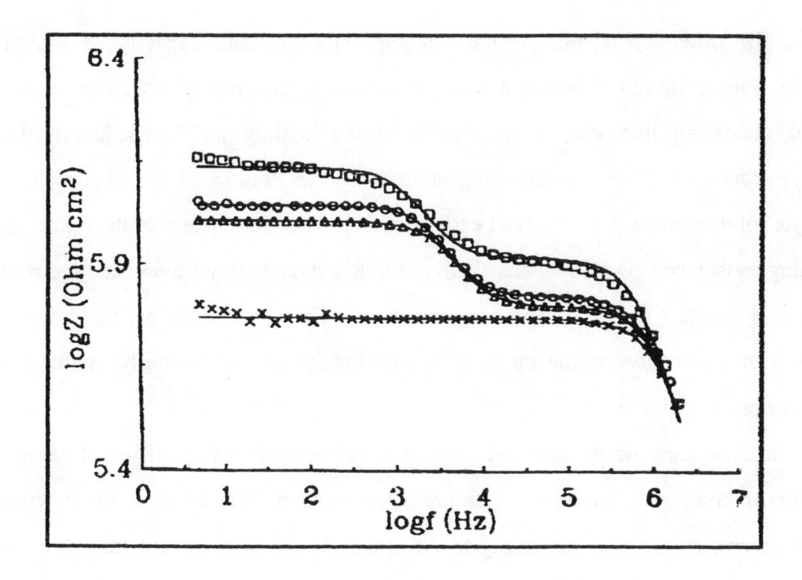

Figure 11 *Bode plots for different types of coated steel specimens exposed to 3.5% NaCl solution for 4h: x, pure epoxy; Δ, 7.5%; O, 15%; □, 30% wt./wt. iron powder.*

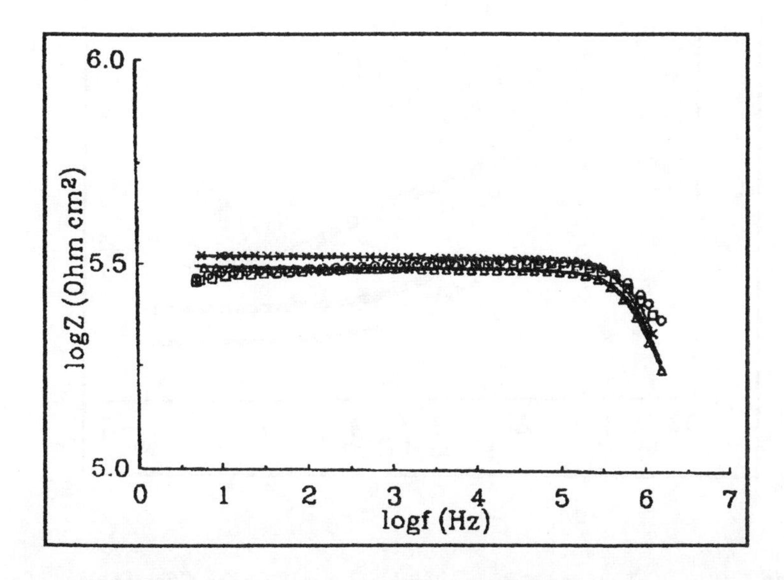

Figure 12 *Bode plots for different types of steel specimens exposed to 3.5% NaCl solution for 90 days: x, pure epoxy; Δ, 7.5%; O, 15%; □, 30% wt./wt. iron powder.*

In the time dependence of the coating resistance and capacitance shown in Figure 13, during the initial period a sharp reduction of the coating resistance is clearly observed indicating increasing conductivity in the coating and consequently lower protective properties. These results suggest that the penetration of the coating by the electrolyte solution increases rapidly at early immersion times. Thereafter the reduction of the coating resistance slows down and finally at long exposure times a tendency to reach a more or less constant resistance value is evidenced. These results reveal that composite coatings with iron concentration up to 30% w/w behave almost similarly as those with pure epoxy resin.

Dielectric permittivity and dielectric loss also provide a way of investigating the behaviour of coatings. Dielectric measurements were performed in a three terminal guarded system, constructed according to ASTM D150-92, with specimens which were taken out of the corrosive environment and formed the dielectric element of the capacitor.

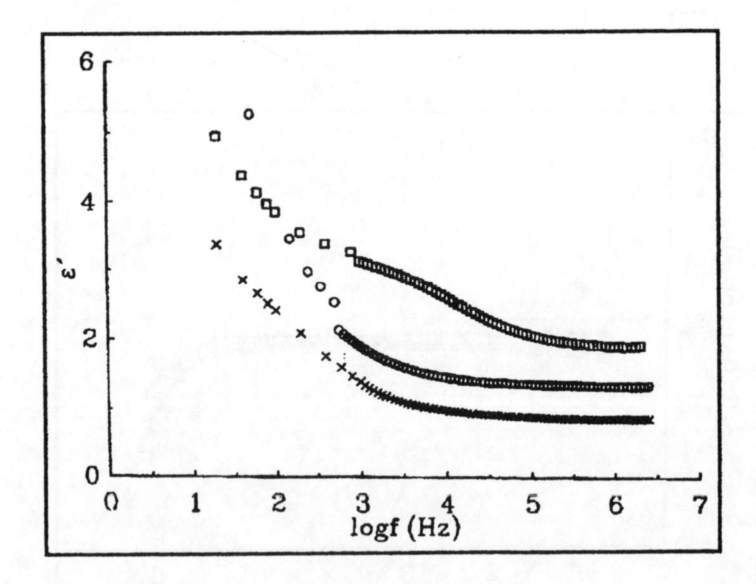

Figure 13 *Change of coating resistance with exposure time for different types of coated steel specimens: x, pure epoxy; Δ, 7.5%; O, 15%; □, 30% wt./wt. iron powder*

As expected (Figure 14) dielectric permittivity is higher for the coatings containing metal powder, here iron powder than for those with pure epoxy resin. Dielectric permittivity is also increasing with the amount of iron powder in the coatings.

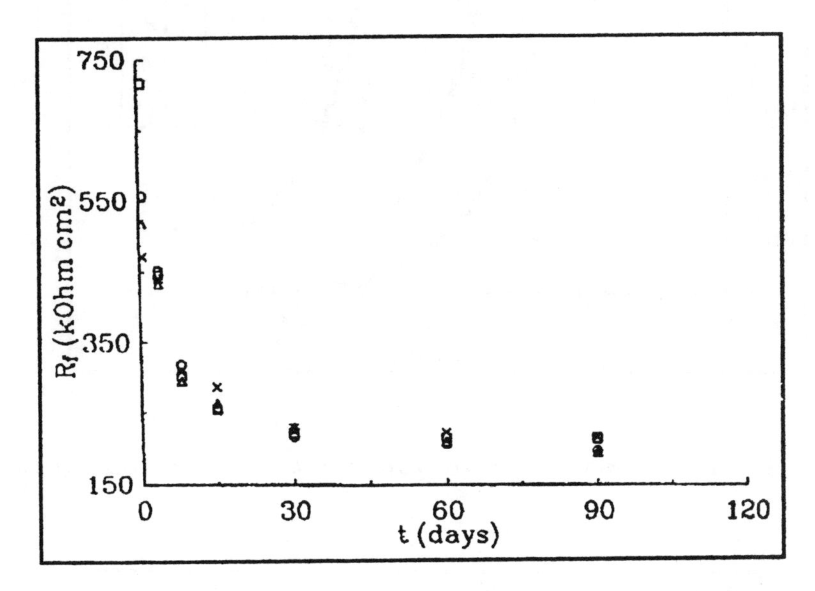

Figure 14 *Dielectric permittivity vs frequency for different types of epoxy coated steel specimens after one day of immersion: x, pure epoxy; Δ, 7.5%; O, 15%; □,*

In the low frequency range dielectric permittivity shows high values which are diminished as the frequency is increased, a behaviour which can be attributed to the interfacial polarisation effect[12,13]. This effect is attributed to heterogeneity of the coatings which is related to the interfaces created by the dispersed particles in the insulating matrix, by the water penetrated and the corrosion products created in the coating.

All coatings show considerable dielectric losses in the low and middle range of frequencies as expected because of interfacial polarisation. As immersion time increases and water uptake is increased in the coatings dielectric losses increase and move to higher frequencies (Figures 15, 16). The straight lines at low frequencies reveal the existence of conductive pathways[14] but since the slope in not substantially altered as immersion time increases, the conductivity of the coating is not considerably changing, which means that the protective behaviour is still existing. Almost the same behaviour was observed for all coatings irrespective of the loading in metal powder.

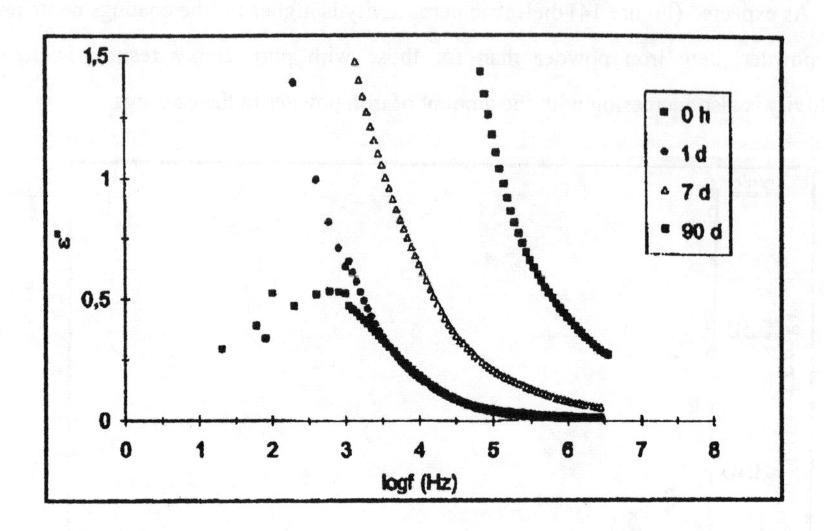

Figure 15 *Dielectric loss vs frequency of epoxy resin coatings at various immersion times*

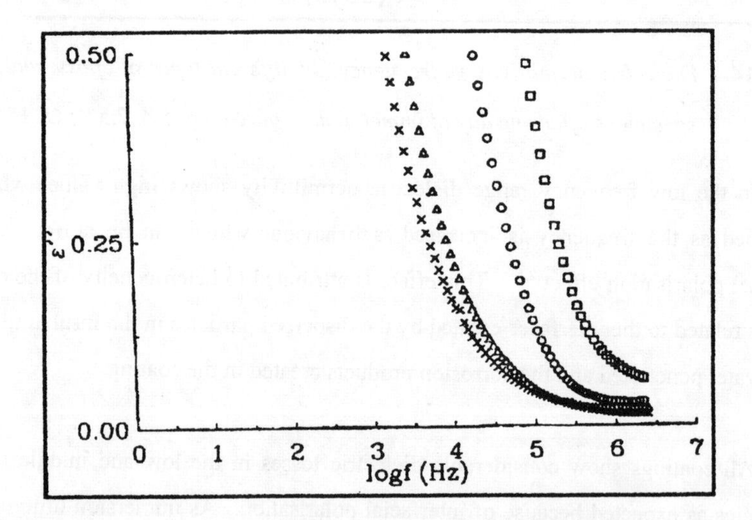

Figure 16 *Dielectric loss vs frequency for the coating with 15% wt/wt. iron powder for different times of immersion: Δ, 1; O, 15.; □, 60 days. The pure epoxy coating after 1 day of immersion is represented by x.*

The almost same slope of the line representing the losses of the pure epoxy coating with the line of losses of coatings containing Fe, Al or Zn powder, also ascertains what it is already found by other measurements that Fe, Al or Zn, in the coating at least, do not worsen the protective properties of the coating.

4. CONCLUSIONS

The particulate composite coatings with Fe, Al and Zn powder provide systems with similar anticorrosive behaviour in chloride environment as the coatings of pure epoxy resin at least for the contents of metal fillers used. The slight increase in the permittivity of these particulate composites is counterbalanced by the barrier effect these particles and their corrosion products exert.

Nickel or copper powders in coatings with epoxy resin reduce corrosion protection performance of the epoxy coatings though they continue to be protective.

References

1. P.A. Schweitzer, "Corrosion and Corrosion Protection Handbook", Marcell Dekker, New York, 1983, p. 375.
2. M.G. Fontana and N.D. Greene "Corrosion Engineering", McGraw-Hill, New York, 2nd edn., 1978, p. 188.
3. B. Ellis, "Chemistry and Technology of Epoxy Resins", Blackie, London, 1993, p.216.
4. B. Normad, A. Pierre and J. Pagetti, in J.M. Costa and D.A. Mocer (eds),Progress in the Understanding and Prevention of Corrosion Vol. 1, Institute of Materials, London 1993, p. 149.
5. M.M. Schwartz, "Composite Materials Handbook", McGraw-Hill, New York 1984, p.376.
6. D.J. Frydrych, G.C. Farrington and H.E. Tewnsend, in M.W. Kending and J. Leidheiser (eds), "Corrosion Protection by Organic Coatings", Electrochemical Society NJ 1987, p. 240.

7. N. Kouloumbi, GM. Tsangaris and S. T. Kybelidis, *J. Coat. Technol. 66*, 1994, **839** 83.
8. N. Kouloumbi, G.M. Tsangaris, A. Skordos, P. Karkanas and J. Kyriopoulou, *Mat. Sci. Forum* 1995, **192-194**, 813.
9. M. Oda, Y. Shibara, T. Yamamoto and T. Morita "Corrosion Protection by Organic Coatings" Ed. M.W. Kending, J. Leidheiser 2, 240, Elec. Soc. (1987).
10. U. Steinsmo and E. Bardal, *J. Electrochem. Soc.* 1984, **136**, 12, 3588.
11. J.N. Murey and H.P.Hack, *Corrosion,* 1991, **47**(6) 480.
12. R.W. Sillars, *J. Inst. Electr. Eng.* 1937, **80**, 378.
13. P. Hedvig, "Dielectric Spectroscopy of Polymers" Adam Hilger, Bristol, 1977 p.282.
14. H. Leidheiser, Jr., *Corrosion NACE*, 1983, **39**(5) 189.

Subject Index